ISBN: 978-1-387-39575-0
ID: 22006094
www.lulu.com

Contents

Preface

Recall from [Berinde, V., Approximation fixed points of weak contractions using the Picard iteration, Nonlinear Analysis Forum, 9 (2004), No. 1, 43-53] that if (X, d) is a metric space, a map $T : X \mapsto X$ is called an almost contraction if there exists $\delta \in [0, 1)$ and $L \geq 0$ such that

$$d(Tx, Ty) \leq \delta d(x, y) + L d(y, Tx)$$

for all $x, y \in X$. Observe that if $L = 0$, then T is a Banach contraction, and by the Banach contraction mapping theorem, T has a unique fixed point. By fixing $L = 1 - \delta$ and $\delta \in (0, 1)$, one says a map $T : X \mapsto X$ is a $(\delta, 1 - \delta)$-weak contraction mapping [Clement Boateng Ampadu, An Almost Contraction Mapping Theorem in Metric Spaces with Unique Fixed Point, Submitted] if for all $x, y \in X$ the following holds

$$d(Tx, Ty) \leq \delta d(x, y) + (1 - \delta) d(y, Tx)$$

It was observed in [Clement Boateng Ampadu, Unique Fixed Point Theorem for Weakly $(\delta, 1 - \delta)$-weak Contractive Mappings, Unpublished] that if $T : X \mapsto X$ is a $(\delta, 1 - \delta)$-weak contraction mapping, then it is nonexpansive when for all $x, y \in X$, the following inequality holds

$$d(Tx, Ty) \leq \frac{1}{2}[d(x, y) + d(y, Tx)]$$

The concepts of total asymptotically nonexpansive mappings [Ya. I. Alber, C. E. Chidume, and H. Zegeye, "Approximating fixed points of total asymptotically nonexpansive mappings," Fixed Point Theory and Applications, vol. 2006, Article ID 10673, 20 pages, 2006], I-asymptotically quasi-nonexpansive mappings [S. Temir, O. Gul, Convergence theorem for I-asymptotically quasi-nonexpansive mapping in Hilbert space, J. Math. Anal. Appl.329 (2007) 759–765], nonself asymptotically I-nonexpansive mappings [L. Yang, X. Xie, Weak and strong convergence theorems for a finite family of I-asymptotically nonexpansive mappings, Appl. Math. Comput. 216 (2010), 1057–1064]; nonself asymptotically nonexpansive mappings [C.E. Chidume, E.U. Ofoedu, H. Zegeye, Strong and weak convergence theorems for asymptotically nonexpansive mappings, J. Math. Anal. Appl., 280 (2003), 364-374], are motivated by when the Banach contraction is nonexpansive.

Motivated by this observation, and when the $(\delta, 1 - \delta)$-weak contraction is nonexpansive, we introduce total asymptotically $(\delta, 1 - \delta)$ nonexpansive mappings, I-asymptotically quasi-$(\delta, 1 - \delta)$-nonexpansive mappings, nonself asymptotically I-$(\delta, 1 - \delta)$-nonexpansive mappings; nonself asymptotically $(\delta, 1 - \delta)$ nonexpansive mappings, and then obtain some convergence theorems of some iterative schemes for a finite family of these mappings in real Banach space (Chapter 1, Chapter 3), convex metric space (Chapter 2), and real uniformly convex Banach space (Chapter 3).

A nice feature of this monograph are the (publishable) exercise set, which begs the reader to explore the beautiful connection between the concepts of total asymptotically nonexpansive mappings, I-asymptotically quasi-nonexpansive mappings, nonself asymptotically I-nonexpansive mappings; nonself asymptotically nonexpansive mappings, which are inspired by when the Banach contraction is nonexpansive, and when the $(\delta, 1 - \delta)$-weak contraction is nonexpansive.

Dedication

This work is dedicated to the nuclear and extended family, friends, colleagues and other people who have shown interest in loving me.

Clement Boateng Ampadu
November, 2017

Chapter 1

Approximation and Convergence of Common Fixed Points of Finite Families of Total Asymptotically $(\delta, 1 - \delta)$ Non-Expansive Mappings

1.1 Brief Summary

Abstract A.1 1

In [Clement Boateng Ampadu, Unique Fixed Point Theorem for Weakly $(\delta, 1 - \delta)$-weak Contractive Mappings, Unpublished] we mentioned when the $(\delta, 1 - \delta)$-weak contraction introduced in [Clement Boateng Ampadu, An Almost Contraction Mapping Theorem in Metric Spaces with Unique Fixed Point, Submitted] is non-expansive. Motivated by certain results contained in [C. E. Chidume, and E. U. Ofoedu, A New Iteration Process for Approximation of Common Fixed Points for Finite Families of Total Asymptotically Nonexpansive Mappings, International Journal of Mathematics and Mathematical Sciences Volume 2009, Article ID 615107, 17 pages] we introduce a concept of total asymptotically $(\delta, 1 - \delta)$ nonexpansive mappings in a real Banach space, E. We then construct a simple iterative sequence in E, and give necessary and sufficient conditions for this sequence to converge to a common fixed point of $\{T_i\}_{i=1}^{m}$, where each T_i for $i = 1, 2, \cdots, m$ is a total asymptotically $(\delta, 1 - \delta)$ nonexpansive mapping defined on a closed convex nonempty subset K of a real Banach space E

1.2 Introduction and Preliminaries

Recall from [C. E. Chidume, and E. U. Ofoedu, A New Iteration Process for Approximation of Common Fixed Points for Finite Families of Total Asymptotically Nonexpansive Mappings, International Journal of Mathematics and Mathematical Sciences Volume 2009, Article ID 615107, 17 pages] that if K is a nonempty subset of a normed real linear space E. A map $T : K \mapsto K$ is said to be nonexpansive if $\|Tx - Ty\| \leq \|x - y\|$ for all $x, y \in K$.

Let (X, d) be a metric space. The concept of $(\delta, 1-\delta)$-weak contraction was introduced in [Clement Boateng Ampadu, An Almost Contraction Mapping Theorem in Metric Spaces with Unique Fixed Point, Submitted] as follows. A map $T : X \mapsto X$ is called a $(\delta, 1 - \delta)$-weak contraction if there exists $\delta \in (0, 1)$ such that $d(Tx, Ty) \leq \delta d(x, y) + (1 - \delta)d(y, Tx)$ for all $x, y \in X$. Observe that

$$d(Tx, Ty) \leq \delta d(x, y) + (1 - \delta)d(y, Tx)$$

is equivalent to

$$d(Tx, Ty) \leq (\delta + 1 - \delta) \max\{d(x, y), d(y, Tx)\}$$

from which we have the following in a normed real linear space E

Definition A.1 1

Let K be a nonempty subset of a normed real linear space E, and $T : E \mapsto E$ be a $(\delta, 1 - \delta)$-weak contraction. $T : K \mapsto K$ will be called $(\delta, 1 - \delta)$-nonexpansive if

$$\|Tx - Ty\| \leq \frac{1}{2}\big[\|x - y\| + \|y - Tx\|\big]$$

for all $x, y \in K$

Recall from [C. E. Chidume, and E. U. Ofoedu, A New Iteration Process for Approximation of Common Fixed Points for Finite Families of Total Asymptotically Nonexpansive Mappings, International Journal of Mathematics and Mathematical Sciences Volume 2009, Article ID 615107, 17 pages] that a nonexpansive mapping is called asymptotically nonexpansive if there exists a sequence $\{\mu_n\}_{n \geq 1} \subset [0, \infty)$ with $\lim_{n \to \infty} \mu_n = 0$ such that for all $x, y \in K$, $\|T^n x - T^n y\| \leq (1 + \mu_n)\|x - y\|$ for all $n \geq 1$. Combining this observation with Definition A.1, we introduce the following

Definition A.2 1

Let K be a nonempty subset of a normed real linear space E, and $T : E \mapsto E$ be a $(\delta, 1 - \delta)$-weak contraction. $T : K \mapsto K$ will be called asymptotically $(\delta, 1 - \delta)$-nonexpansive if

$$\|T^n x - T^n y\| \leq (\frac{1}{2} + \mu_n)\big[\|x - y\| + \|y - T^n x\|\big]$$

for all $x, y \in K$ and $n \geq 1$, where $\{\mu_n\}_{n \geq 1} \subset [0, \infty)$ is a sequence with $\lim_{n \to \infty} \mu_n = 0$

Recall from [C. E. Chidume, and E. U. Ofoedu, A New Iteration Process for Approximation of Common Fixed Points for Finite Families of Total Asymptotically Nonexpansive Mappings, International Journal of Mathematics and Mathematical Sciences Volume 2009, Article ID 615107, 17 pages] that a nonexpansive mapping is called uniformly L-Lipschitzian if there exists a constant $L \geq 0$ such that for all $x, y \in K$, $\|T^n x - T^n y\| \leq L\|x - y\|$ for all $n \geq 1$. Combining this observation with Definition A.2, we introduce the following

Definition A.3 1

Let K be a nonempty subset of a normed real linear space E, and $T : E \mapsto E$ be a $(\delta, 1 - \delta)$-weak contraction. $T : K \mapsto K$ will be called uniformly L-Lpschitzian if there exists a constant $L \geq 0$ such that

$$\|T^n x - T^n y\| \leq L\big[\|x - y\| + \|y - T^n x\|\big]$$

for all $x, y \in K$ and $n \geq 1$.

Note that the class of asymptotically nonexpansive mappings was introduced by Goebel and Kirk [K. Goebel and W. A. Kirk, "A fixed point theorem for asymptotically nonexpansive mappings," Proceedings of the American Mathematical Society, vol. 35, pp. 171–174, 1972] as a generalization of the class of nonexpansive mappings. They proved that if K is a nonempty closed convex bounded subset of a uniformly convex real Banach space and T is an asymptotically nonexpansive self-mapping of K, then T has a fixed point. Clearly, we have the following

Question A.4 1

If K is a nonempty closed convex bounded subset of a uniformly convex real Banach space and T is an asymptotically $(\delta, 1 - \delta)$-nonexpansive self-mapping of K. What can we say about the fixed point of T?

Recall from [C. E. Chidume, and E. U. Ofoedu, A New Iteration Process for Approximation of Common Fixed Points for Finite Families of Total Asymptotically Nonexpansive Mappings, International Journal of Mathematics and Mathematical Sciences Volume 2009, Article ID 615107,

17 pages] that a mapping T is said to be asymptotically nonexpansive in the intermediate sense if it is continuous and the following inequality holds

$$\limsup_{n \to \infty} \sup_{x,y \in K} (\|T^n x - T^n y\| - \|x - y\|) \leq 0$$

The above definition motivates the following

Definition A.5 1

Let K be a nonempty subset of a normed real linear space E, and $T : E \mapsto E$ be a $(\delta, 1-\delta)$-weak contraction. $T : K \mapsto K$ will be called asymptotically $(\delta, 1 - \delta)$ nonexpansive in the intermediate sense if it is continuous and the following inequality holds

$$\limsup_{n \to \infty} \sup_{x,y \in K} \left(\|T^n x - T^n y\| - \frac{1}{2} [\|x - y\| + \|y - T^n x\|] \right) \leq 0$$

Observe if we put

$$a_n := \sup_{x,y \in K} \left(\|T^n x - T^n y\| - \frac{1}{2} [\|x - y\| + \|y - T^n x\|] \right)$$

and

$$\sigma_n = \max\{0, a_n\}$$

then $\lim_{n \to \infty} \sigma_n = 0$ and so the inequality in Definition A.5 immediately above reduces to

$$\|T^n x - T^n y\| \leq \frac{1}{2} [\|x - y\| + \|y - T^n x\|] + \sigma_n$$

for all $x, y \in K$, $n \geq 1$.

Note that the class of mappings which are asymptotically nonexpansive in the intermediate sense was introduced by Bruck et al. [R. Bruck, T. Kuczumow, and S. Reich, "Convergence of iterates of asymptotically nonexpansive mappings in Banach spaces with the uniform Opial property," Colloquium Mathematicum, vol. 65, no. 2, pp. 169–179, 1993]. It is known [W. A. Kirk, "Fixed point theorems for non-Lipschitzian mappings of asymptotically nonexpansive type," Israel Journal of Mathematics, vol. 17, no. 4, pp. 339–346, 1974] that if K is a nonempty closed convex bounded subset of a uniformly convex real Banach space E and T is a self-mapping of K which is asymptotically nonexpansive in the intermediate sense, then T has a fixed point. Clearly we have the following

---

Question A.6 1

If K is a nonempty closed convex bounded subset of a uniformly convex real Banach space E and T is a self-mapping of K which is asymptotically $(\delta, 1 - \delta)$ nonexpansive in the intermediate sense. What can we say about the fixed point of T?

Recall from [C. E. Chidume, and E. U. Ofoedu, A New Iteration Process for Approximation of Common Fixed Points for Finite Families of Total Asymptotically Nonexpansive Mappings, International Journal of Mathematics and Mathematical Sciences Volume 2009, Article ID 615107, 17 pages] that if K is a nonempty subset of a normed space E and $\{a_n\}_{n \geq 1}$ is a sequence in $[0, +\infty)$ such that $\lim_{n \to \infty} a_n = 0$, then $T : K \mapsto K$ is called nearly Lipschitzian with respect to $\{a_n\}_{n \geq 1}$ if for each $n \in \mathbb{N}$ there exists $k_n \geq 0$ such that $\|T^n x - T^n y\| \leq k_n(\|x - y\| + a_n)$ for all $x, y \in K$. It appears the class of nearly Lipschitzian mappings were introduced in [D. R. Sahu, "Fixed points of demicontinuous nearly Lipschitzian mappings in Banach spaces," Commentationes Mathematicae Universitatis Carolinae, vol. 46, no. 4, pp. 653–666, 2005]. This definition motivates the following

CHAPTER 1. **APPROXIMATION AND CONVERGENCE OF COMMON FIXED POINTS OF FINITE FAMILIES OF TOTAL ASYMPTOTICALLY** $(\delta, 1 - \delta)$ **NON-EXPANSIVE MAPPINGS**

8

Definition A.7 1

Let K be a nonempty subset of a normed space E, and $T : E \mapsto E$ be a $(\delta, 1 - \delta)$-weak contraction. $T : K \mapsto K$ will be called nearly Lipschitzian with respect to $\{a_n\}_{n \geq 1}$ a sequence in $[0, +\infty)$ with $\lim_{n \to \infty} a_n = 0$, if for each $n \in \mathbb{N}$ there exists $k_n \geq 0$ such that

$$\|T^n x - T^n y\| \leq k_n \big[\|x - y\| + \|y - T^n x\| + a_n\big]$$

for all $x, y \in K$. Moreover, the nearly Lipschitz constant is defined as

$$\alpha(T^n) := \sup \left\{ \frac{\|T^n x - T^n y\|}{[\|x - y\| + \|y - T^n x\| + a_n]} : x, y \in K, x \neq y \right\}$$

and satisfies the following

$$\|T^n x - T^n y\| \leq \alpha(T^n)\big[\|x - y\| + \|y - T^n x\| + a_n\big]$$

for all $x, y \in K$, $n \in \mathbb{N}$. In particular, we say

(a) T is nearly $(\delta, 1 - \delta)$ nonexpansive if $\alpha(T^n) = \frac{1}{2}$ for all $n \in \mathbb{N}$

(b) T is nearly asymptotically $(\delta, 1 - \delta)$ nonexpansive if $\alpha(T^n) \geq \frac{1}{2}$ for all $n \in \mathbb{N}$ and $\lim_{n \to \infty} \alpha(T^n) = \frac{1}{2}$

(c) T is nearly uniform L-Lipschitzian if $\alpha(T^n) \leq L$ for all $n \in \mathbb{N}$

Example A.8 1

Let $E = \mathbb{R}$, $K = [0, 1]$. Define $T : K \mapsto K$ by $Tx = \frac{1}{2}$ if $x \in [0, \frac{1}{2}]$; $Tx = 0$ if $x \in (\frac{1}{2}, 1]$. Clearly T is not continuous, and hence is not $(\delta, 1 - \delta)$-weak Lipschitz, that is, for some $L \geq 0$, it does not satisfy

$$\|Tx - Ty\| \leq L\big[\|x - y\| + \|y - Tx\|\big]$$

for all $x, y \in K$. However, observe that T is nearly $(\delta, 1 - \delta)$ nonexpansive. In fact, take a real sequence $\{a_n\}_{n \geq 1}$ with $a_1 = \frac{1}{2}$ and $\lim_{n \to \infty} a_n = 0$, then,

$$\|T^n x - T^n y\| \leq \frac{1}{2}\big[\|x - y\| + \|y - T^n x\| + a_1\big]$$

for all $x, y \in K$, and

$$\|T^n x - T^n y\| \leq \frac{1}{2}\big[\|x - y\| + \|y - T^n x\| + a_n\big]$$

for all $x, y \in K$ and $n \geq 2$. This is because $T^n x = \frac{1}{2}$ for all $x \in [0, 1]$, $n \geq 2$

Observe that if K is a bounded domain of an asymptotically $(\delta, 1 - \delta)$-nonexpansive mapping T, then T is nearly asymptotically $(\delta, 1 - \delta)$ nonexpansive. In fact for all $x, y \in K$ and $n \in \mathbb{N}$, we have,

$$\|T^n x - T^n y\| \leq (\frac{1}{2} + \mu_n)\big[\|x - y\| + \|y - T^n x\|\big]$$

$$\leq \frac{1}{2}\big[\|x - y\| + \|y - T^n x\|\big] + \mu_n\big[\|x - y\| + \|y - T^n x\|\big]$$

$$\leq \frac{1}{2}\big[\|x - y\| + \|y - T^n x\|\big] + diam(K)\mu_n$$

Observe that if K is a bounded domain of a nearly asymptotically $(\delta, 1 - \delta)$ nonexpansive mapping T, then T is asymptotically $(\delta, 1 - \delta)$ nonexpansive in the intermediate sense. To see this, let T be a nearly asymptotically $(\delta, 1 - \delta)$ nonexpansive mapping, then,

$$\|T^n x - T^n y\| \leq \alpha(T^n)\big[\|x - y\| + \|y - T^n x\| + a_n\big]$$

for all $x, y \in K$, $n \in \mathbb{N}$. Now observe that

$$\|T^n x - T^n y\| - \frac{1}{2}\|x - y\| \leq \alpha(T^n)\big[\|x - y\| + \|y - T^n x\| + a_n\big] - \frac{1}{2}\|x - y\|$$

$$\leq \big(\alpha(T^n) - \frac{1}{2}\big)\|x - y\| + \alpha(T^n)\|y - T^n x\| + \alpha(T^n)a_n$$

From the above, we deduce that

$$\|T^n x - T^n y\| - \frac{1}{2}\big[\|x - y\| + \|y - T^n x\|\big] \leq \big(\alpha(T^n) - \frac{1}{2}\big)\big[\|x - y\| + \|y - T^n x\|\big] + \alpha(T^n)a_n$$

It then follows that for all $n \geq 1$, we have

$$\sup_{x,y \in K}\left(\|T^n x - T^n y\| - \frac{1}{2}\big[\|x - y\| + \|y - T^n x\|\big]\right) \leq \big(\alpha(T^n) - \frac{1}{2}\big)diam(K) + \alpha(T^n)a_n$$

Hence,

$$\limsup_{n \to \infty}\sup_{x,y \in K}\left(\|T^n x - T^n y\| - \frac{1}{2}\big[\|x - y\| + \|y - T^n x\|\big]\right) \leq 0$$

The main tool for approximation of fixed points of generalizations of nonexpansive mappings remains iterative technique (see, e.g, Ref. [6-19, 20,22,23,24,25,26, and 28] contained in [C. E. Chidume, and E. U. Ofoedu, A New Iteration Process for Approximation of Common Fixed Points for Finite Families of Total Asymptotically Nonexpansive Mappings, International Journal of Mathematics and Mathematical Sciences Volume 2009, Article ID 615107, 17 pages]).

It is proved in [S. C. Bose, "Weak convergence to the fixed point of an asymptotically nonexpansive map," Proceedings of the American Mathematical Society, vol. 68, no. 3, pp. 305–308, 1978] that if K is a nonempty closed convex bounded subset of a uniformly convex real Banach space E satisfying Opial's condition [Z. Opial, "Weak convergence of the sequence of successive approximations for nonexpansive mappings," Bulletin of the American Mathematical Society, vol. 73, no. 4, pp. 591–597, 1967] (i.e., for all sequences $\{x_n\}$ in E such that $\{x_n\}$ converges weakly to some $x \in E$, the inequality $\liminf_{n \to \infty}\|x_n - x\| < \liminf_{n \to \infty}\|x_n - y\|$ holds for all $y \neq x$ in E) and $T : K \mapsto K$ is an asymptotically nonexpansive mapping, then the sequence $\{T^n x\}$ converges weakly to a fixed point of T provided that T is asymptotically regular at $x \in K$, that is, the limit, $\lim_{n \to \infty}\|T^n x - T^{n+1} x\| = 0$, holds. Clearly, we have the following

> **Question A.9 1**
>
> Suppose that K is a nonempty closed convex bounded subset of a uniformly convex real Banach space E satisfying Opial's condition [Z. Opial, "Weak convergence of the sequence of successive approximations for nonexpansive mappings," Bulletin of the American Mathematical Society, vol. 73, no. 4, pp. 591–597, 1967] (i.e., for all sequences $\{x_n\}$ in E such that $\{x_n\}$ converges weakly to some $x \in E$, the inequality $\liminf_{n \to \infty}\|x_n - x\| < \liminf_{n \to \infty}\|x_n - y\|$ holds for all $y \neq x$ in E) and $T : K \mapsto K$ is an asymptotically $(\delta, 1 - \delta)$ nonexpansive mapping, then under what conditions does the sequence $\{T^n x\}$ converge weakly to a fixed point of T?

Recall from [C. E. Chidume, and E. U. Ofoedu, A New Iteration Process for Approximation of Common Fixed Points for Finite Families of Total Asymptotically Nonexpansive Mappings, International Journal of Mathematics and Mathematical Sciences Volume 2009, Article ID 615107, 17 pages] that a map $T : K \mapsto K$ is called totally asymptotically nonexpansive (which appears was first introduced in [Ya. I. Alber, C. E. Chidume, and H. Zegeye, "Approximating fixed points of total asymptotically nonexpansive mappings," Fixed Point Theory and Applications, vol. 2006, Article ID 10673, 20 pages, 2006]) if there exists nonnegative real sequences $\{\mu_n\}$ and $\{l_n\}$, $n \geq 1$ with $\mu_n \to 0$ and $l_n \to 0$ as $n \to \infty$ and a strictly increasing continuous function $\phi : \mathbb{R}^+ \mapsto \mathbb{R}^+$ with $\phi(0) = 0$ such that for all $x, y \in K$ and $n \geq 1$ the following inequality holds

$$\|T^n x - T^n y\| \leq \|x - y\| + \mu_n \phi(\|x - y\|) + l_n$$

> **Definition A.10 1**
>
> $T : K \mapsto K$ will be called total asymptotically $(\delta, 1-\delta)$ nonexpansive if there exists nonnegative real sequences $\{\mu_n\}$ and $\{l_n\}$, $n \geq 1$ with $\mu_n \to 0$ and $l_n \to 0$ as $n \to \infty$ and a strictly increasing continuous function $\phi : \mathbb{R}^+ \mapsto \mathbb{R}^+$ with $\phi(0) = 0$ such that for all $x, y \in K$ and $n \geq 1$ the following inequality holds
>
> $$\|T^n x - T^n y\| \leq \frac{1}{2}\big[\|x - y\| + \|y - T^n x\|\big] + \mu_n \phi\big(\big[\|x - y\| + \|y - T^n x\|\big]\big) + l_n$$

Observe that if $\phi(\lambda) = \lambda$, then, the inequality in the previous definition reduces to

$$\|T^n x - T^n y\| \leq (\frac{1}{2} + \mu_n)\big[\|x - y\| + \|y - T^n x\|\big] + l_n$$

If $l_n = 0$ for all $n \geq 1$ in the inequality immediately above, then total asymptotically $(\delta, 1-\delta)$ nonexpansive mappings coincide with asymptotically $(\delta, 1-\delta)$ nonexpansive mappings.

If we take $\mu_n = 0$ and $l_n = 0$ for all $n \geq 1$ in the inequality of Definition A.10, then we obtain the class of mappings that includes the class of $(\delta, 1-\delta)$ nonexpansive mappings. If $\mu_n = 0$, and $l_n = \sigma_n = \max\{0, a_n\}$, where

$$a_n := \sup_{x,y \in K}\left(\|T^n x - T^n y\| - \frac{1}{2}\big[\|x - y\| + \|y - T^n x\|\big]\right)$$

for all $n \geq 1$, then the inequality in Definition A.10 reduces to the equivalent form of the inequality in Definition A.5.

Recall from [C. E. Chidume, and E. U. Ofoedu, A New Iteration Process for Approximation of Common Fixed Points for Finite Families of Total Asymptotically Nonexpansive Mappings, International Journal of Mathematics and Mathematical Sciences Volume 2009, Article ID 615107, 17 pages] that another class of nonlinear mappings introduced as a further generalization of nonexpansive mappings with nonempty fixed point sets is the class of asymptotically quasinonexpansive mappings which properly contains the class of asymptotically nonexpansive operators with nonempty fixed point sets (e.g. see Ref. [8,16,30-33] contained in [C. E. Chidume, and E. U. Ofoedu, A New Iteration Process for Approximation of Common Fixed Points for Finite Families of Total Asymptotically Nonexpansive Mappings, International Journal of Mathematics and Mathematical Sciences Volume 2009, Article ID 615107, 17 pages]). Moreover, a mapping T is called quasi-nonexpansive if $F(T) \neq \emptyset$ and $\|Tx - x^*\| \leq \|x - x^*\|$, for all $x \in D(T)$, $x^* \in F(T)$. This definition motivates the following

> **Definition A.11 1**
>
> Let T be a $(\delta, 1-\delta)$ nonexpansive mapping. T will be called quasi-$(\delta, 1-\delta)$-nonexpansive if $F(T) \neq \emptyset$ and
>
> $$\|Tx - x^*\| \leq \frac{1}{2}\big[\|x - x^*\| + \|x^* - Tx\|\big]$$
>
> for all $x \in D(T)$, $x^* \in F(T)$

Recall from [C. E. Chidume, and E. U. Ofoedu, A New Iteration Process for Approximation of Common Fixed Points for Finite Families of Total Asymptotically Nonexpansive Mappings, International Journal of Mathematics and Mathematical Sciences Volume 2009, Article ID 615107, 17 pages] that a nonexpansive mapping is called asymptotically quasi-nonexpansive if $F(T) \neq \emptyset$ and there exists a sequence $\{\mu_n\} \in [0, \infty)$ with $\lim_{n \to \infty} \mu_n = 0$ such that for all $x \in D(T)$ and $x^* \in F(T)$, one has for $n \geq 1$, $\|T^n x - x^*\| \leq (1 + \mu_n)\|x - x^*\|$. This definition motivates the following

Definition A.12 1

Let T be a $(\delta, 1 - \delta)$ nonexpansive mapping. T will be called asymptotically quasi-$(\delta, 1 - \delta)$-nonexpansive if $F(T) \neq \emptyset$ and there exists a sequence $\{\mu_n\} \in [0, \infty)$ with $\lim_{n \to \infty} \mu_n = 0$ such that for all $x \in D(T)$ and $x^* \in F(T)$, one has for $n \geq 1$,

$$\|T^n x - x^*\| \leq \left(\frac{1}{2} + \mu_n\right)\left[\|x - x^*\| + \|x^* - T^n x\|\right]$$

Recall from [C. E. Chidume, and E. U. Ofoedu, A New Iteration Process for Approximation of Common Fixed Points for Finite Families of Total Asymptotically Nonexpansive Mappings, International Journal of Mathematics and Mathematical Sciences Volume 2009, Article ID 615107, 17 pages] that a mapping T is said to be asymptotically quasi-nonexpansive in the intermediate sense if it is continuous and the following inequality holds

$$\limsup_{n \to \infty} \left\{ \sup_{x \in D(T), x^* \in F(T)} (\|T^n x - x^*\| - \|x - x^*\|) \right\} \leq 0$$

This definition motivates the following

Definition A.13 1

Let T be a $(\delta, 1 - \delta)$ nonexpansive mapping. T will be called asymptotically quasi-$(\delta, 1 - \delta)$-nonexpansive in the intermediate sense if it is continuous and the following inequality holds

$$\limsup_{n \to \infty} \left\{ \sup_{x \in D(T), x^* \in F(T)} \left(\|T^n x - x^*\| - \frac{1}{2}\left[\|x - x^*\| + \|x^* - T^n x\|\right] \right) \right\} \leq 0$$

Observe if we put

$$a_n^* := \sup_{x \in D(T), x^* \in F(T)} \left(\|T^n x - x^*\| - \frac{1}{2}\left[\|x - x^*\| + \|x^* - T^n x\|\right] \right)$$

and

$$\sigma_n^* = \max\{0, a_n^*\}$$

then $\lim_{n \to \infty} \sigma_n^* = 0$ and so the inequality in the definition immediately above reduces to

$$\|T^n x - x^*\| \leq \frac{1}{2}\left[\|x - x^*\| + \|x^* - T^n x\|\right] + \sigma_n^*$$

for all $x, y \in K$, $n \geq 1$.

Existence theorems for common fixed points of certain families of nonlinear mappings have been established by various authors (see, e.g. Ref. [2,34-37] contained in [C. E. Chidume, and E. U. Ofoedu, A New Iteration Process for Approximation of Common Fixed Points for Finite Families of Total Asymptotically Nonexpansive Mappings, International Journal of Mathematics and Mathematical Sciences Volume 2009, Article ID 615107, 17 pages]). Also considerable research efforts have been devoted to developing iterative methods for approximating common fixed points when they exist of finite families of this class of mappings (see, e.g. Ref. [16, 33, 38-46] contained in [C. E. Chidume, and E. U. Ofoedu, A New Iteration Process for Approximation of Common Fixed Points for Finite Families of Total Asymptotically Nonexpansive Mappings, International Journal of Mathematics and Mathematical Sciences Volume 2009, Article ID 615107, 17 pages]).

Chidume and Ofoedu [C. E. Chidume and E. U. Ofoedu, "Approximation of common fixed points for finite families of total asymptotically nonexpansive mappings," Journal of Mathematical Analysis and Applications, vol. 333, no. 1, pp. 128–141, 2007] introduced an iterative scheme for approximation of a common fixed point of a finite family of total asymptotically nonexpansive mappings in Banach spaces. More precisely, they proved the following theorems

CHAPTER 1. **APPROXIMATION AND CONVERGENCE OF COMMON FIXED POINTS OF FINITE FAMILIES OF TOTAL ASYMPTOTICALLY** $(\delta, 1 - \delta)$ **NON-EXPANSIVE MAPPINGS**

12

Theorem A.14 1

[C. E. Chidume and E. U. Ofoedu, "Approximation of common fixed points for finite families of total asymptotically nonexpansive mappings," Journal of Mathematical Analysis and Applications, vol. 333, no. 1, pp. 128–141, 2007] Let E be a real Banach space, K be a nonempty closed convex subset of E, and $T_i : K \mapsto K$, $i = 1, 2, \cdots, m$ be m total asymptotically nonexpansive mappings with sequences $\{\mu_{in}\}$, $\{l_{in}\}$, $n \geq 1$, $i = 1, 2, \cdots, m$ such that $F := \bigcap_{i=1}^{m} F(T_i) \neq \emptyset$. Let $\{x_n\}$ be given by

$$x_1 \in E,$$

$$x_{n+1} = (1 - \alpha_n)x_n + \alpha_n T_1^n x_n, \ if \ m = 1, \ n \geq 1,$$

$$x_1 \in E,$$

$$x_{n+1} = (1 - \alpha_n)x_n + \alpha_n T_1^n y_{1n}$$

$$y_{1n} = (1 - \alpha_n)x_n + \alpha_n T_2^n y_{2n}$$

$$\vdots$$

$$y_{(m-2)n} = (1 - \alpha_n)x_n + \alpha_n T_{m-1}^n y_{(m-1)n}$$

$$y_{(m-1)n} = (1 - \alpha_n)x_n + \alpha_n T_m^n x_n, \ if \ m \geq 2, \ n \geq 1,$$

Suppose $\sum_{n=1}^{\infty} \mu_{in} < \infty$, $\sum_{n=1}^{\infty} l_{in} < \infty$, $i = 1, 2, \cdots, m$, and suppose there exists $M_i, M_i^* > 0$ such that $\phi_i(\lambda_i) \leq M_i^* \lambda_i$ for all $\lambda_i \geq M_i$, $i = 1, 2, \cdots, m$. Then the sequence $\{x_n\}$ is bounded and $\lim_{n \to \infty} \|x_n - p\|$ exists, $p \in F$. Moreover, the sequence $\{x_n\}$ converges strongly to a common fixed point of T_i, $i = 1, 2, \cdots, m$ iff $\liminf_{n \to \infty} d(x_n, F) = 0$, where $d(x_n, F) = \inf_{y \in F} \|x_n - y\|$, $n \geq 1$

Theorem A.15 1

[C. E. Chidume and E. U. Ofoedu, "Approximation of common fixed points for finite families of total asymptotically nonexpansive mappings," Journal of Mathematical Analysis and Applications, vol. 333, no. 1, pp. 128–141, 2007] Let E be a uniformly convex real Banach space, and K be a nonempty closed convex subset of E, and $T_i : K \mapsto K$, $i = 1, 2, \cdots, m$ be m uniformly continuous total asymptotically nonexpansive mappings with sequences $\{\mu_{in}\}, \{l_{in}\} \in [0, \infty)$ such that $\sum_{n=1}^{\infty} \mu_{in} < \infty$, $\sum_{n=1}^{\infty} l_{in} < \infty$, $i = 1, 2, \cdots, m$ and $F := \bigcap_{i=1}^{m} F(T_i) \neq \emptyset$. Let $\{\alpha_{in}\} \subset [\epsilon, 1 - \epsilon]$ for some $\epsilon \in (0, 1)$. From arbitrary $x_1 \in E$, let $\{x_n\}$ be given by the previous theorem. Suppose there exists $M_i, M_i^* > 0$ such that $\phi_i(\lambda_i) \leq M_i^* \lambda_i$ for all $\lambda_i \geq M_i$, $i = 1, 2, \cdots, m$, and that one of T_i for $i = 1, 2, \cdots, m$ is compact, then $\{x_n\}$ converges strongly to some $p \in F$

In the next section we construct a new iterative sequence much simpler than that contained in Theorem A.14 for approximation of common fixed points of finite families of total asymptotically $(\delta, 1 - \delta)$ nonexpansive mappings and give necessary and sufficient conditions for the convergence of the scheme to common fixed points of the mappings in arbitrary real Banach spaces.

In the sequel we will need the following

Lemma A.16 1

[C. E. Chidume, and E. U. Ofoedu, A New Iteration Process for Approximation of Common Fixed Points for Finite Families of Total Asymptotically Nonexpansive Mappings, International Journal of Mathematics and Mathematical Sciences Volume 2009, Article ID 615107, 17 pages] Let $\{a_n\}$, $\{\alpha_n\}$, and $\{b_n\}$ be sequences of nonnegative real numbers such that

$$a_{n+1} \leq (1 + \alpha_n)a_n + b_n$$

Suppose that $\sum_{n=1}^{\infty} \alpha_n < \infty$ and $\sum_{n=1}^{\infty} b_n < \infty$. Then $\{a_n\}$ is bounded and $\lim_{n \to \infty} a_n$ exists. Moreover, if in addition $\liminf_{n \to \infty} a_n = 0$, then, $\lim_{n \to \infty} a_n = 0$.

1.3 Main Results

Let K be a nonempty closed convex subset of a real normed space E. Let $T_1, T_2, \cdots, T_m : K \mapsto K$ be m total asymptotically $(\delta, 1 - \delta)$ nonexpansive mappings. We define the iterative sequence $\{x_n\}$ by

$$x_1 \in K$$

$$x_{n+1} = (1 - \alpha_{0n})x_n + \sum_{i=1}^{m} \alpha_{in} T_i^n x_n, \ n \geq 1$$

where $\{\alpha_{in}\}_{n=1}^{\infty}$, $i = 0, 1, 2, \cdots, m$ are sequences in $(0, 1)$ such that $\sum_{i=0}^{m} \alpha_{in} = 1$. Our first main result is the following

> **Theorem A.1 1**
>
> Let E be a real Banach space, K be a nonempty closed convex subset of E, and $T_i : K \mapsto K$, $i = 1, 2, \cdots, m$ be m total asymptotically $(\delta, 1 - \delta)$ nonexpansive mappings with sequences $\{\mu_{in}\}, \{l_{in}\}$, $n \geq 1$, $i = 1, 2, \cdots, m$ such that $F := \bigcap_{i=1}^{m} F(T_i) \neq \emptyset$. Let $\{x_n\}$ be given by
>
> $$x_1 \in K$$
>
> $$x_{n+1} = (1 - \alpha_{0n})x_n + \sum_{i=1}^{m} \alpha_{in} T_i^n x_n, \ n \geq 1$$
>
> where $\{\alpha_{in}\}_{n=1}^{\infty}$, $i = 0, 1, 2, \cdots, m$ are sequences in $(0, 1)$ such that $\sum_{i=0}^{m} \alpha_{in} = 1$. Suppose $\sum_{n=1}^{\infty} \mu_{in} < \infty$, $\sum_{n=1}^{\infty} l_{in} < \infty$, $i = 1, 2, \cdots, m$, and suppose there exists $M_i, M_i^* > 0$ such that $\phi_i(\lambda_i) \leq M_i^* \lambda_i$ for all $\lambda_i \geq M_i$, $i = 1, 2, \cdots, m$. Then the sequence $\{x_n\}$ is bounded and $\lim_{n \to \infty} \|x_n - p\|$ exists, $p \in F$.

Proof of Theorem A.1 1

Observe that since $\sum_{i=0}^{m} \alpha_{in} = 1$, we have, $\sum_{i=1}^{n} \alpha_{in} \leq 1 - \alpha_{0n}$. Thus,

$$2 \sum_{i=1}^{n} \alpha_{in} \leq 1 - \alpha_{0n} + \sum_{i=1}^{n} \alpha_{in}$$

$$\leq 1 - (\alpha_{0n} - \sum_{i=1}^{n} \alpha_{in})$$

$$< 1$$

Thus, $\sum_{i=1}^{n} \alpha_{in} \leq \frac{1}{2}$

Let $p \in F$. Then from the definition of $\{x_n\}$ we have

$$\|x_{n+1} - p\| = \|(1 - \alpha_{0n})x_n + \sum_{i=1}^{m} \alpha_{in} T_i^n x_n - p\|$$

$$= \|(\sum_{i=1}^{m} \alpha_{in})x_n + \sum_{i=1}^{m} \alpha_{in} T_i^n x_n - p\|$$

$$\leq \|(1 - \sum_{i=1}^{m} \alpha_{in})x_n + \sum_{i=1}^{m} \alpha_{in} T_i^n x_n - p\|$$

$$= \|(1 - \sum_{i=1}^{m} \alpha_{in})(x_n - p) + \sum_{i=1}^{m} \alpha_{in}(T_i^n x_n - p)\|$$

$$\leq (1 - \sum_{i=1}^{m} \alpha_{in})\|x_n - p\| + \sum_{i=1}^{m} \alpha_{in}\|T_i^n x_n - p\|$$

$$\leq (1 - \sum_{i=1}^{m} \alpha_{in})\|x_n - p\| + \sum_{i=1}^{m} \alpha_{in}\left\{\frac{1}{2}\big[\|x_n - p\| + \|p - T^n x_n\|\big]\right.$$

$$\left. + \mu_{in}\phi_i\big([\|x_n - p\| + \|p - T^n x_n\|]\big) + l_{in}\right\}$$

Since ϕ_i is an increasing function, it follows that $\phi_i(\lambda_i) \leq \phi_i(M_i)$ whenever $\lambda_i \leq M_i$ and (by hypothesis) $\phi_i(\lambda_i) \leq M_i^* \lambda_i$ if $\lambda_i \geq M_i$. In either case, we have

$$\phi_i\big([\|x_n - p\| + \|p - T^n x_n\|]\big) \leq \phi_i(M_i) + M_i^*\big[\|x_n - p\| + \|p - T^n x_n\|\big]$$

for some $M_i, M_i^* > 0$. Thus, from the chain of inequalities immediately above, we deduce that

$$\|x_{n+1} - p\| \leq (1 - \sum_{i=1}^{m} \alpha_{in})\|x_n - p\| + \sum_{i=1}^{m} \alpha_{in}\left\{\frac{1}{2}\big[\|x_n - p\| + \|p - T^n x_n\|\big]\right.$$

$$\left. + \mu_{in}\phi_i(M_i) + \mu_{in}M_i^*\big[\|x_n - p\| + \|p - T^n x_n\|\big] + \mu_{in}l_{in}\right\}$$

From the above we deduce that

Proof of Theorem A.1 Continued 1

$$\|x_{n+1} - p\| \leq \left(1 - \sum_{i=1}^{m} \alpha_{in} + \frac{1}{2}\sum_{i=1}^{m}\alpha_{in}\right.$$

$$\left. + \sum_{i=1}^{m} \mu_{in}M_i^*\alpha_{in}\right)\|x_n - p\| + \left(\sum_{i=1}^{m}\alpha_{in} + \sum_{i=1}^{m}\mu_{in}M_i^*\alpha_{in}\right)\|p - T^n x_n\|$$

$$+ \sum_{i=1}^{m}\alpha_{in}\mu_{in}\phi_i(M_i) + \sum_{i=1}^{m}\alpha_{in}\mu_{in}l_{in}$$

$$\leq \left(1 + \frac{1}{2}\sum_{i=1}^{m}\alpha_{in} + 2\sum_{i=1}^{m}\mu_{in}M_i^*\alpha_{in}\right)\|x_n - p\| + \sum_{i=1}^{m}\alpha_{in}\mu_{in}\phi_i(M_i) + \sum_{i=1}^{m}\alpha_{in}\mu_{in}l_{in}$$

In order to invoke Lemma A.16 in the above inequality, put

$$\alpha_n := \frac{1}{2}\sum_{i=1}^{m}\alpha_{in} + 2\sum_{i=1}^{m}\mu_{in}M_i^*\alpha_{in}$$

and

$$b_n := \sum_{i=1}^{m}\alpha_{in}\mu_{in}\phi_i(M_i) + \sum_{i=1}^{m}\alpha_{in}\mu_{in}l_{in}$$

Observe that $\sum_{n=1}^{\infty}\alpha_n < \infty$ and $\sum_{n=1}^{\infty} b_n < \infty$. So by Lemma 1.16, it follows that the sequence $\{x_n\}$ is bounded and that $\lim_{n\to\infty}\|x_n - p\|$ exists

Now we give necessary and sufficient conditions for convergence in real Banach spaces.

Theorem A.2 1

Let E be a real Banach space, and let K be a nonempty closed convex subset of E, and let $T_i : K \mapsto K$, $i = 1, 2, \cdots, m$ be m continuous total asymptotically $(\delta, 1 - \delta)$ nonexpansive mappings with sequences $\{\mu_{in}\}$, $\{l_{in}\}$, $n \geq 1$, $i = 1, 2, \cdots, m$ such that $F := \bigcap_{i=1}^{m} F(T_i) \neq \emptyset$. Let $\{x_n\}$ be defined as in the previous theorem. Suppose $\sum_{n=1}^{\infty}\mu_{in} < \infty$, $\sum_{n=1}^{\infty} l_{in} < \infty$, $i = 1, 2, \cdots, m$, and suppose there exists $M_i, M_i^* > 0$ such that $\phi_i(\lambda_i) \leq M_i^*\lambda_i$ for all $\lambda_i \geq M_i$, $i = 1, 2, \cdots, m$. Then the sequence $\{x_n\}$ converges strongly to a common fixed point of T_i, $i = 1, 2, \cdots, m$ iff $\liminf_{n\to\infty} d(x_n, F) = 0$, where $d(x_n, F) = \inf_{y\in F}\|x_n - y\|$, $n \geq 1$

Proof of Theorem A.2 1

It suffices to show that $\liminf_{n \to \infty} d(x_n, F) = 0$ implies that $\{x_n\}$ converges to a common fixed point of T_i, $i = 1, 2, \cdots, m$.

Necessity: From the previous theorem we know that $\|x_{n+1} - p\| \leq (1 + \alpha_n)\|x_n - p\| + b_n$ holds for all $n \geq 1$ with

$$\alpha_n := \frac{1}{2} \sum_{i=1}^{m} \alpha_{in} + 2 \sum_{i=1}^{m} \mu_{in} M_i^* \alpha_{in}$$

and

$$b_n := \sum_{i=1}^{m} \alpha_{in} \mu_{in} \phi_i(M_i) + \sum_{i=1}^{m} \alpha_{in} \mu_{in} l_{in}$$

Since it also holds for all $p \in F$, we obtain from it that for all $n \geq 1$

$$d(x_{n+1}, F) \leq (1 + \alpha_n)d(x_n, F) + b_n$$

Lemma A.16 implies that $\lim_{n \to \infty} d(x_n, F)$ exists, but $\liminf_{n \to \infty} d(x_n, F) = 0$. Hence, $\lim_{n \to \infty} d(x_n, F) = 0$.

Sufficiency: We first show that $\{x_n\}$ is a Cauchy sequence in E. For all integers $m \geq 1$, we obtain from the necessity portion of the proof that

$$\|x_{n+m} - p\| \leq \prod_{i=n}^{n+m-1} (1 + \alpha_i)\|x_n - p\| + \left(\sum_{i=n}^{n+m-1} b_i \right) \prod_{i=n}^{n+m-1} (1 + \alpha_i)$$

$$\leq exp\left(\sum_{i=n}^{n+m-1} \alpha_i \right)\|x_n - p\| + \left(\sum_{i=n}^{n+m-1} b_i \right) exp\left(\sum_{i=n}^{n+m-1} \alpha_i \right)$$

It follows that for all integers $m \geq 1$ and all $p \in F$, one has

$$\|x_{n+m} - x_n\| \leq \|x_{n+m} - p\| + \|x_n - p\|$$

$$\leq \left[1 + exp\left(\sum_{i=n}^{n+m-1} \alpha_i \right) \right]\|x_n - p\| + \left(\sum_{i=n}^{n+m-1} b_i \right) exp\left(\sum_{i=n}^{n+m-1} \alpha_i \right)$$

From the above, it follows that for some constant $D > 0$, we have

$$\|x_{n+m} - x_n\| \leq (1 + D)\|x_n - p\| + \left(\sum_{i=n}^{\infty} b_i \right) D$$

Taking infimum over $p \in F$ in the previous inequality, we have

$$\|x_{n+m} - x_n\| \leq (1 + D)d(x_n, F) + \left(\sum_{i=n}^{\infty} b_i \right) D$$

Now since $\lim_{n \to \infty} d(x_n, F) = 0$ and $\sum_{i=n}^{\infty} b_i < \infty$, given $\epsilon > 0$, there exists an integer $N_1 > 0$ such that for all $n \geq N_1$, $d(x_n, F) < \frac{\epsilon}{2(D+1)}$, and $\sum_{i=n}^{\infty} b_i < \frac{\epsilon}{2D}$. It follows that for all integers $n \geq N_1$, $m \geq 1$, we have, $\|x_{n+m} - x_n\| < \epsilon$. Hence $\{x_n\}$ is a Cauchy sequence in E, and since E is complete there exists $l^* \in E$ such that $\lim_{n \to \infty} x_n = l^*$. Now we show that l^* is a common fixed point of T_i, $i = 1, 2, \cdots, m$, that is, we show that $l^* \in F$. Suppose for contradiction that $l^* \in F^c$ (where F^c denotes the complement of F). Since F is a closed subset of E (recall each T_i, $i = 1, 2, \cdots, m$ is continuous), we have that $d(l^*, F) > 0$.

Proof of Theorem A.2 Continued 1

However, for all $p \in F$, we have,

$$\|l^* - p\| \leq \|l^* - x_n\| + \|x_n - p\|$$

which implies that

$$d(l^*, F) \leq \|x_n - l^*\| + d(x_n, F)$$

If we take limits as $n \to \infty$ in the inequality immediately above, we obtain $d(l^*, F) = 0$ which contradicts $d(l^*, F) > 0$. Thus, l^* is a common fixed point of T_i, $i = 1, 2, \cdots, m$

If $T_1, \cdots, T_m$ are asymptotically $(\delta, 1 - \delta)$ nonexpansive mappings, then $l_{in} = 0$ for all $n \geq 1$, $i = 1, 2, \cdots, m$ and $\phi(\lambda_i) = \lambda_i$ so the assumption that there exists $M_i, M_i^* > 0$ such that $\phi(\lambda_i) \leq M_i^* \lambda_i$ for all $\lambda_i \geq M_i$, $i = 1, 2, \cdots, m$ in the above theorems is no longer needed. In particular we have the following

Corollary A.3 1

Let E be a real Banach space, and K be a nonempty closed convex subset of E, and let $T_i : K \mapsto K$, $i = 1, 2, \cdots, m$ be m continuous asymptotically $(\delta, 1 - \delta)$ nonexpansive mappings with sequences $\{\mu_{in}\}$, $n \geq 1$, $i = 1, 2, \cdots, m$ such that $F := \bigcap_{i=1}^{m} F(T_i) \neq \emptyset$. Let $\{x_n\}$ be defined as in Theorem A.1. Suppose that $\sum_{n=1}^{\infty} \mu_{in} < \infty$, $i = 1, 2, \cdots, m$. Then the sequence $\{x_n\}$ is bounded and $\lim_{n \to \infty} \|x_n - p\|$ exists, $p \in F$. Moreover, $\{x_n\}$ converges strongly to a common fixed point of T_i, $i = 1, 2, \cdots, m$ if and only if $\liminf_{n \to \infty} d(x_n, F) = 0$

1.4 Exercises

Recall from [Clement Boateng Ampadu, A Strong Ciric Almost Contraction Mapping Theorem in Metric Spaces with Unique Fixed Point, Applied Mathematics(AM), Initially Accepted but Withdrawn] that if (X, d) is a metric space, then a map $T : X \mapsto X$ is called a strong Ciric $(\alpha, 1 - \alpha)$-weak contraction if there exists $\alpha \in (0, 1)$ such that for all $x, y \in X$ the following holds

$$d(Tx, Ty) \leq \alpha M_1(x, y) + (1 - \alpha) d(y, Tx)$$

where

$$M_1(x, y) = \max\{d(x, y), d(x, Tx), d(y, Ty), \frac{1}{2}[d(x, Ty) + d(y, Tx)]\}$$

Observe

$$d(Tx, Ty) \leq \alpha M_1(x, y) + (1 - \alpha) d(y, Tx)$$

where

$$M_1(x, y) = \max\{d(x, y), d(x, Tx), d(y, Ty), \frac{1}{2}[d(x, Ty) + d(y, Tx)]\}$$

is equivalent to

$$d(Tx, Ty) \leq (\alpha + 1 - \alpha) \max\{M_1(x, y), d(y, Tx)\}$$

where

$$M_1(x, y) = \max\{d(x, y), d(x, Tx), d(y, Ty), \frac{1}{2}[d(x, Ty) + d(y, Tx)]\}$$

and so we have the following in a normed real linear space E

Let K be a nonempty subset of a normed real linear space E, and $T : E \mapsto E$ be a strong Ciric $(\alpha, 1 - \alpha)$-weak contraction. $T : K \mapsto K$ will be called $(\alpha, 1 - \alpha)$-nonexpansive if for all $x, y \in K$ the following holds

$$\|Tx - Ty\| \leq \frac{1}{2}[M_1(x, y) + \|y - Tx\|]$$

where

$$M_1(x, y) = \max\{\|x - y\|, \|x - Tx\|, \|y - Ty\|, \frac{1}{2}[\|x - Ty\| + \|y - Tx\|]\}$$

Also, we say $T : K \mapsto K$ is total asymptotically $(\alpha, 1 - \alpha)$ nonexpansive if there exists nonnegative real sequences $\{\mu_n\}$ and $\{l_n\}$, $n \geq 1$ with $\mu_n \to 0$ and $l_n \to 0$ as $n \to \infty$ and a strictly increasing continuous function $\phi : \mathbb{R}^+ \mapsto \mathbb{R}^+$ with $\phi(0) = 0$ such that for all $x, y \in K$ and $n \geq 1$ the following inequality holds

$$\|T^n x - T^n y\| \leq \frac{1}{2}[M_2(x, y) + \|y - T^n x\|] + \mu_n \phi([M_2(x, y) + \|y - T^n x\|]) + l_n$$

where

$$M_2(x, y) = \max\{\|x - y\|, \|x - T^n x\|, \|y - T^n y\|, \frac{1}{2}[\|x - T^n y\| + \|y - T^n x\|]\}$$

Exercise A.1 1

Let E be a real Banach space, K be a nonempty closed convex subset of E, and $T_i : K \mapsto K$, $i = 1, 2, \cdots, m$ be m total asymptotically $(\alpha, 1 - \alpha)$ nonexpansive mappings with sequences $\{\mu_{in}\}$, $\{l_{in}\}$, $n \geq 1$, $i = 1, 2, \cdots, m$ such that $F := \bigcap_{i=1}^{m} F(T_i) \neq \emptyset$. Let $\{x_n\}$ be given by

$$x_1 \in K$$

$$x_{n+1} = (1 - \alpha_{0n})x_n + \sum_{i=1}^{m} \alpha_{in} T_i^n x_n, \ n \geq 1$$

where $\{\alpha_{in}\}_{n=1}^{\infty}$, $i = 0, 1, 2, \cdots, m$ are sequences in $(0, 1)$ such that $\sum_{i=0}^{m} \alpha_{in} = 1$. Suppose $\sum_{n=1}^{\infty} \mu_{in} < \infty$, $\sum_{n=1}^{\infty} l_{in} < \infty$, $i = 1, 2, \cdots, m$, and suppose there exists $M_i, M_i^* > 0$ such that $\phi_i(\lambda_i) \leq M_i^* \lambda_i$ for all $\lambda_i \geq M_i$, $i = 1, 2, \cdots, m$. Then the sequence $\{x_n\}$ is bounded and $\lim_{n \to \infty} \|x_n - p\|$ exists, $p \in F$.

Exercise A.2 1

Let E be a real Banach space, and let K be a nonempty closed convex subset of E, and let $T_i : K \mapsto K$, $i = 1, 2, \cdots, m$ be m continuous total asymptotically $(\alpha, 1 - \alpha)$ nonexpansive mappings with sequences $\{\mu_{in}\}$, $\{l_{in}\}$, $n \geq 1$, $i = 1, 2, \cdots, m$ such that $F := \bigcap_{i=1}^{m} F(T_i) \neq \emptyset$. Let $\{x_n\}$ be defined as in the previous exercise. Suppose $\sum_{n=1}^{\infty} \mu_{in} < \infty$, $\sum_{n=1}^{\infty} l_{in} < \infty$, $i = 1, 2, \cdots, m$, and suppose there exists $M_i, M_i^* > 0$ such that $\phi_i(\lambda_i) \leq M_i^* \lambda_i$ for all $\lambda_i \geq M_i$, $i = 1, 2, \cdots, m$. Then the sequence $\{x_n\}$ converges strongly to a common fixed point of T_i, $i = 1, 2, \cdots, m$ iff $\liminf_{n \to \infty} d(x_n, F) = 0$, where $d(x_n, F) = \inf_{y \in F} \|x_n - y\|$, $n \geq 1$

1.5 References

(1) Clement Boateng Ampadu, Unique Fixed Point Theorem for Weakly $(\delta, 1 - \delta)$-weak Contractive Mappings, Unpublished. Available online: https://drive.google.com/file/d/0BwtkpMtWoUlEdXVNQVBGUVRWbEU/view

(2) Clement Boateng Ampadu, An Almost Contraction Mapping Theorem in Metric Spaces with Unique Fixed Point, Submitted. Available online: https://drive.google.com/file/d/0BwtkpMtWoUlEY25OZW1HUEdGcU0/view

(3) C. E. Chidume, and E. U. Ofoedu, A New Iteration Process for Approximation of Common Fixed Points for Finite Families of Total Asymptotically Nonexpansive Mappings, International Journal of Mathematics and Mathematical Sciences Volume 2009, Article ID 615107, 17 pages

(4) K. Goebel and W. A. Kirk, "A fixed point theorem for asymptotically nonexpansive mappings," Proceedings of the American Mathematical Society, vol. 35, pp. 171–174, 1972

(5) R. Bruck, T. Kuczumow, and S. Reich, "Convergence of iterates of asymptotically nonexpansive mappings in Banach spaces with the uniform Opial property," Colloquium Mathematicum, vol. 65, no. 2, pp. 169–179, 1993

(6) W. A. Kirk, "Fixed point theorems for non-Lipschitzian mappings of asymptotically nonexpansive type," Israel Journal of Mathematics, vol. 17, no. 4, pp. 339–346,1974

(7) D. R. Sahu, "Fixed points of demicontinuous nearly Lipschitzian mappings in Banach spaces," Commentationes Mathematicae Universitatis Carolinae, vol. 46, no. 4, pp. 653–666, 2005

(8) S. C. Bose, "Weak convergence to the fixed point of an asymptotically nonexpansive map," Proceedings of the American Mathematical Society, vol. 68, no. 3, pp. 305–308, 1978

(9) Z. Opial, "Weak convergence of the sequence of successive approximations for nonexpansive mappings," Bulletin of the American Mathematical Society, vol. 73, no. 4, pp. 591–597, 1967

(10) Ya. I. Alber, C. E. Chidume, and H. Zegeye, "Approximating fixed points of total asymptotically nonexpansive mappings," Fixed Point Theory and Applications, vol. 2006, Article ID 10673, 20 pages, 2006

(11) C. E. Chidume and E. U. Ofoedu, "Approximation of common fixed points for finite families of total asymptotically nonexpansive mappings," Journal of Mathematical Analysis and Applications, vol. 333, no. 1, pp. 128–141, 2007

(12) Clement Boateng Ampadu, A Strong Ciric Almost Contraction Mapping Theorem in Metric Spaces with Unique Fixed Point, Applied Mathematics(AM), Initially Accepted but Withdrawn. Available online:
https://drive.google.com/file/d/0BwtkpMtWoUlEZHhpdExKVGdMOTA/view

Chapter 2

Approximating Common Fixed Points of a Finite Family of I-Asymptotically and Asymptotically Quasi-$(\delta, 1 - \delta)$-Non Expansive Mappings

2.1 Brief Summary

Proof of Lemma B.2 Continued 1

In [Clement Boateng Ampadu, Unique Fixed Point Theorem for Weakly $(\delta, 1 - \delta)$-weak Contractive Mappings, Unpublished] we mentioned when the $(\delta, 1 - \delta)$-weak contraction introduced in [Clement Boateng Ampadu, An Almost Contraction Mapping Theorem in Metric Spaces with Unique Fixed Point, Submitted] is non-expansive. Motivated by certain results contained in [Birol Gunduz, Fixed Points of a Finite Family of I-Asymptotically Quasi-Nonexpansive Mappings in a Convex Metric Space, Filomat 31:7 (2017), 2175–2182] we introduce a concept of I-asymptotically quasi-$(\delta, 1 - \delta)$-nonexpansive mappings in a convex metric space, and study the Ishikawa iterative scheme with error terms for this class of mappings. We establish strong convergence theorems with application for the proposed algorithms in a convex metric space.

2.2 Preliminaries

> **Notation B.1 1**
>
> Throughout this paper
>
> (a) $\mathbb{N}$ will denote the set of natural numbers
>
> (b) $J = \{1, 2, \cdots, r\}$ will denote the set of first r natural numbers
>
> (c) $F(T)$ will denote the set of fixed points of T
>
> (d) Let $\{T_i : i \in J\}$ and $\{I_i : i \in J\}$ be two finite families of mappings.
>
> $$F := \left(\bigcap_{i=1}^{r} F(T_i) \right) \cap \left(\bigcap_{i=1}^{r} F(I_i) \right)$$
>
> will denote the set of common fixed points of $\{T_i : i \in J\}$ and $\{I_i : i \in J\}$

Let (X, d) be a metric space, and $T : X \mapsto X$ be a mapping. Recall from [Birol Gunduz, Fixed Points of a Finite Family of I-Asymptotically Quasi-Nonexpansive Mappings in a Convex Metric Space, Filomat 31:7 (2017), 2175–2182] that a mapping T is called nonexpansive if for all $x, y \in X$, we have, $d(Tx, Ty) \leq d(x, y)$.

On the other hand the concept of $(\delta, 1 - \delta)$-weak contraction was introduced in [Clement Boateng Ampadu, An Almost Contraction Mapping Theorem in Metric Spaces with Unique Fixed Point, Submitted] as follows. A map $T : X \mapsto X$ is called a $(\delta, 1 - \delta)$-weak contraction if there exists $\delta \in (0, 1)$ such that $d(Tx, Ty) \leq \delta d(x, y) + (1 - \delta) d(y, Tx)$ for all $x, y \in X$. Observe that

$$d(Tx, Ty) \leq \delta d(x, y) + (1 - \delta) d(y, Tx)$$

is equivalent to

$$d(Tx, Ty) \leq (\delta + 1 - \delta) \max\{d(x, y), d(y, Tx)\}$$

from which we have the following

> **Definition B.2 1**
>
> Let (X, d) be a metric space, and $T : X \mapsto X$ be a $(\delta, 1 - \delta)$-weak contraction. T will be called nonexpansive if for all $x, y \in X$, we have,
>
> $$d(Tx, Ty) \leq \frac{1}{2}[d(x, y) + d(y, Tx)]$$

Recall from [Birol Gunduz, Fixed Points of a Finite Family of I-Asymptotically Quasi-Nonexpansive Mappings in a Convex Metric Space, Filomat 31:7 (2017), 2175–2182] that T is called quasi-nonexpansive if $F(T) \neq \emptyset$ and $d(Tx, p) \leq d(x, p)$ for all $x \in X$ and $p \in F(T)$. Now we introduce the following

> **Definition B.3 1**
>
> Let (X, d) be a metric space, and $T : X \mapsto X$ be a $(\delta, 1 - \delta)$-weak contraction. T will be called quasi-$(\delta, 1 - \delta)$-nonexpansive if $F(T) \neq \emptyset$, and for all $x \in X$ and $p \in F(T)$, we have,
>
> $$d(Tx, p) \leq \frac{1}{2}[d(x, p) + d(p, Tx)]$$

Recall from [Birol Gunduz, Fixed Points of a Finite Family of I-Asymptotically Quasi-Nonexpansive Mappings in a Convex Metric Space, Filomat 31:7 (2017), 2175–2182] that T is called asymptotically-nonexpansive if there exists $u_n \in [0, \infty)$ for all $n \in \mathbb{N}$ with $\lim_{n \to \infty} u_n = 0$ such that $d(T^n x, T^n y) \leq (1 + u_n) d(x, y)$ for all $x, y \in X$ and $n \in \mathbb{N}$. Now we introduce the following

> **Definition B.4 1**
>
> Let (X, d) be a metric space, and $T : X \mapsto X$ be a $(\delta, 1 - \delta)$-weak contraction. T will be called asymptotically $(\delta, 1 - \delta)$-nonexpansive if there exists $u_n \in [0, \infty)$ for all $n \in \mathbb{N}$ with $\lim_{n \to \infty} u_n = 0$ such that
>
> $$d(T^n x, T^n y) \leq (\frac{1}{2} + u_n)[d(x, y) + d(y, T^n x)]$$
>
> for all $x, y \in X$ and $n \in \mathbb{N}$

Recall from [Birol Gunduz, Fixed Points of a Finite Family of I-Asymptotically Quasi-Nonexpansive Mappings in a Convex Metric Space, Filomat 31:7 (2017), 2175–2182] that T is called asymptotically quasi-nonexpansive if $F(T) \neq \emptyset$ and there exists $u_n \in [0, \infty)$ for all $n \in \mathbb{N}$ with $\lim_{n \to \infty} u_n = 0$ such that $d(T^n x, p) \leq (1 + u_n)d(x, p)$ for all $x \in X$, $p \in F(T)$ and $n \in \mathbb{N}$. Now we introduce the following

> **Definition B.5 1**
>
> Let (X, d) be a metric space, and $T : X \mapsto X$ be a $(\delta, 1 - \delta)$-weak contraction. T will be called asymptotically quasi-$(\delta, 1 - \delta)$-nonexpansive if $F(T) \neq \emptyset$ and there there exists $u_n \in [0, \infty)$ for all $n \in \mathbb{N}$ with $\lim_{n \to \infty} u_n = 0$ such that
>
> $$d(T^n x, p) \leq (\frac{1}{2} + u_n)[d(x, p) + d(p, T^n x)]$$
>
> for all $x \in X$, $p \in F(T)$, and $n \in \mathbb{N}$

One concept which generalizes the notion of asymptotically nonexpansive mapping in Banach space is *I*-asymptotically nonexpansive mapping defined by Temir and Gul [S. Temir, On the convergence theorems of implicit iteration process for a finite family of I-asymptotically nonexpansive mappings, J. Comput. Appl. Math. 225 (2009) 398–405; S. Temir, O. Gul, Convergence theorem for I-asymptotically quasi-nonexpansive mapping in Hilbert space, J. Math. Anal. Appl.329 (2007) 759–765]. The metric counterpart of these mappings appeared as follows

> **Definition B.6 1**
>
> [Birol Gunduz, Fixed Points of a Finite Family of I-Asymptotically Quasi-Nonexpansive Mappings in a Convex Metric Space, Filomat 31:7 (2017), 2175–2182] Let X be a metric space and $T, I : X \mapsto X$ be two mappings. T is said to be
>
> (a) *I*-asymptotically nonexpansive if there exists a sequence $\{v_n\} \subset [0, \infty)$ with $\lim_{n \to \infty} v_n = 0$ such that
>
> $$d(T^n x, T^n y) \leq (1 + v_n)d(I^n x, I^n y)$$
>
> for all $x, y \in X$ and $n \geq 1$
>
> (b) *I*-asymptotically quasi-nonexpansive if $F(T) \cap F(I) \neq \emptyset$ and there exists a sequence $\{v_n\} \subset [0, \infty)$ with $\lim_{n \to \infty} v_n = 0$ such that
>
> $$d(T^n x, p) \leq (1 + v_n)d(I^n x, p)$$
>
> for all $x \in X$, $p \in F(T) \cap F(I)$ and $n \geq 1$
>
> (c) *I*-uniformly Lipschitz if there exists $\Gamma > 0$ such that
>
> $$d(T^n x, T^n y) \leq \Gamma d(I^n x, I^n y)$$
>
> for all $x, y \in X$ and $n \geq 1$

Now we introduce the following

Definition B.7 1

Let X be a metric space and $T, I : X \mapsto X$ be two mappings. T will be called

(a) *I*-asymptotically $(\delta, 1-\delta)$ nonexpansive if there exists a sequence $\{v_n\} \subset [0, \infty)$ with $\lim_{n \to \infty} v_n = 0$ such that

$$d(T^n x, T^n y) \leq (\frac{1}{2} + v_n)[d(I^n x, I^n y) + d(I^n y, T^n x)]$$

for all $x, y \in X$ and $n \geq 1$

(b) *I*-asymptotically quasi-$(\delta, 1-\delta)$-nonexpansive if $F(T) \cap F(I) \neq \emptyset$ and there exists a sequence $\{v_n\} \subset [0, \infty)$ with $\lim_{n \to \infty} v_n = 0$ such that

$$d(T^n x, p) \leq (\frac{1}{2} + v_n)[d(I^n x, p) + d(p, T^n x)]$$

for all $x \in X$, $p \in F(T) \cap F(I)$ and $n \geq 1$

(c) *I*-uniformly Lipschitz if there exists $\Gamma > 0$ such that

$$d(T^n x, T^n y) \leq \Gamma[d(I^n x, I^n y) + d(I^n y, T^n x)]$$

for all $x, y \in X$ and $n \geq 1$

Definition B.8 1

[W. Takahashi, A convexity in metric space and nonexpansive mappings, Kodai. Math. Sem. Rep. 22 (1970) 142-149] A convex structure in a metric space (X, d) is a mapping $W : X \times X \times [0, 1] \mapsto X$ satisfying, for all $x, y, u \in X$ and all $\lambda \in [0, 1]$

$$d(u, W(x, y; \lambda)) \leq \lambda d(u, x) + (1 - \lambda) d(u, y)$$

A metric space together with a convex structure is called a convex metric space. A nonempty subset C of X is said to be convex if $W(x, y; \lambda) \in C$ for all $(x, y; \lambda) \in C \times C \times [0, 1]$

The above definition was extended as follows

Definition B.9 1

[Birol Gunduz, Fixed Points of a Finite Family of I-Asymptotically Quasi-Nonexpansive Mappings in a Convex Metric Space, Filomat 31:7 (2017), 2175–2182] A convex structure in a metric space (X, d) is a mapping $W : X^3 \times [0, 1]^3 \mapsto X$ satisfying, for any $(x, y, z; a, b, c) \in X^3 \times [0, 1]^3$ with $a + b + c = 1$, and $u \in X$,

$$d(u, W(x, y, z; a, b, c)) \leq ad(u, x) + bd(u, y) + cd(u, z)$$

A metric space together with a convex structure is called a convex metric space. A nonempty subset C of X is said to be convex if $W(x, y, z; a, b, c) \in C$ for all $(x, y, z) \in C^3$, $(a, b, c) \in [0, 1]^3$ with $a + b + c = 1$

In [S. Temir, On the convergence theorems of implicit iteration process for a finite family of I-asymptotically nonexpansive mappings, J. Comput. Appl. Math. 225 (2009) 398–405] introduced an iteration process for a finite family of *I*-asymptotically nonexpansive mappings in Banach space as follows. Let K be a nonempty subset of a Banach space X. Let $\{T_i\}_{i=1}^{N}$ be a finite family of I_i-asymptotically nonexpansive self-mappings and let $\{I_i\}_{i=1}^{N}$ be a finite family of asymptotically nonexpansive self-mappings of K. Let $\{\alpha_n\}$ and $\{\beta_n\}$ be two real sequences in $[0, 1]$. Then the sequence $\{x_n\}$ is generated as follows

$$x_{n+1} = (1 - \alpha_n)x_n + \alpha_n I_{i(n)}^{k(n)} y_n$$

$$y_n = (1 - \beta_n)x_n + \beta_n T_{i(n)}^{k(n)} x_n$$

where $n = (k(n) - 1)N + i(n)$, $i(n) \in \{1, 2, \cdots, N\}$, $n \geq 1$.

The above iteration process was transformed with error terms for a finite family of *I*-asymptotically quasi-nonexpansive mappings in convex metric spaces as follows

Definition B.10 1

[Birol Gunduz, Fixed Points of a Finite Family of I-Asymptotically Quasi-Nonexpansive Mappings in a Convex Metric Space, Filomat 31:7 (2017), 2175–2182] Let (X, d) be a convex metric space with convex structure W, $\{T_i : i \in J\}$ be a finite family of I_i-asymptotically quasi-nonexpansive mappings. Suppose that $\{u_n\}$ and $\{v_n\}$ are two bounded sequences in X and $\{\alpha_n\}$, $\{\beta_n\}$, $\{\gamma_n\}$, $\{\widehat{\alpha}_n\}$, $\{\widehat{\beta}_n\}$, $\{\widehat{\gamma}_n\}$ are six sequences in $[0, 1]$ such that $\alpha_n + \beta_n + \gamma_n = 1 = \widehat{\alpha}_n + \widehat{\beta}_n + \widehat{\gamma}_n$ for $n \in \mathbb{N}$. For any given $x_1 \in X$, $\{x_n\}$ is defined by

$$x_{n+1} = W\left(x_n, I_i^n y_n, u_n; \alpha_n, \beta_n, \gamma_n\right)$$

$$y_n = W\left(x_n, T_i^n x_n, v_n; \widehat{\alpha}_n, \widehat{\beta}_n, \widehat{\gamma}_n\right)$$

where $n = (k - 1)r + i \geq 1$, $i = i(n) \in J$ is a positive integer and $k(n) \to \infty$ as $n \to \infty$

Remark B.11 1

Note that the sequence $\{x_n\}$ given in the previous definition can be defined for $n \geq 1$ as

$$x_{n+1} = W\left(x_n, I_{i(n)}^{k(n)} y_n, u_n; \alpha_n, \beta_n, \gamma_n\right)$$

$$y_n = W\left(x_n, T_{i(n)}^{k(n)} x_n, v_n; \widehat{\alpha}_n, \widehat{\beta}_n, \widehat{\gamma}_n\right)$$

In this paper we use the above definition to prove some strong convergence results for approximating common fixed points of a finite family of *I*-asymptotically quasi-$(\delta, 1 - \delta)$ nonexpansive mappings and a finite family of asymptotically quasi-$(\delta, 1 - \delta)$ nonexpansive mappings in a convex metric space. In what follows we will need the following

Lemma B.12 1

[L. Qihou, Iterative sequence for asymptotically quasi-nonexpansive mappings with errors member, J. Math. Anal. Appl. 259 (2001) 18–24] Let $\{a_n\}$, $\{b_n\}$, and $\{c_n\}$ be three nonnegative sequences satisfying $\sum_{n=0}^{\infty} b_n < \infty$, $\sum_{n=0}^{\infty} c_n < \infty$, and $a_{n+1} = (1 + b_n)a_n + c_n$ for $n \geq 0$, then

(a) $\lim_{n \to \infty} a_n$ exists

(b) if $\liminf_{n \to \infty} a_n = 0$, then $\lim_{n \to \infty} a_n = 0$

2.3 Main Results

Proposition B.1 1

Let (X, d) be a convex metric space with convex structure W, $\{T_i : i \in J\} : X \mapsto X$ be a finite family of I_i-asymptotically quasi-$(\delta, 1-\delta)$-nonexpansive mappings and $\{I_i : i \in J\} : X \mapsto X$ be a finite family of asymptotically quasi-$(\delta, 1-\delta)$ nonexpansive mappings with $F := \left(\bigcap_{i=1}^{r} F(T_i) \right) \cap \left(\bigcap_{i=1}^{r} F(I_i) \right) \neq \emptyset$. Then there exists a point $p \in F$ and sequences $\{k_n\}, \{l_n\} \in [0, \infty)$ with $\lim_{n \to \infty} k_n = \lim_{n \to \infty} l_n = 0$ such that for all $x \in K$ and for each $i \in I$, the following inequalities hold

$$d(T_i^n x, p) \leq (\frac{1}{2} + k_n)[d(I_i^n x, p) + d(p, T_i^n x)]$$

$$d(I_i^n x, p) \leq (\frac{1}{2} + l_n)[d(x, p) + d(p, I_i^n x)]$$

Proof of Proposition B.1 1

Since $\{T_i : i \in J\} : X \mapsto X$ is a finite family of I_i-asymptotically quasi-$(\delta, 1-\delta)$-nonexpansive mappings and $\{I_i : i \in J\} : X \mapsto X$ is a finite family of asymptotically quasi-$(\delta, 1-\delta)$ nonexpansive mappings with $F := \left(\bigcap_{i=1}^{r} F(T_i) \right) \cap \left(\bigcap_{i=1}^{r} F(I_i) \right) \neq \emptyset$, then there exists $p \in F$ and sequences $\{k_{in}\}, \{l_{in}\} \in [0, \infty)$ with $\lim_{n \to \infty} k_{in} = \lim_{n \to \infty} l_{in} = 0$ for each $i \in J$ such that for each $x \in X$, the following inequalities hold

$$d(T_i^n x, p) \leq (\frac{1}{2} + k_{in})[d(I_i^n x, p) + d(p, T_i^n x)]$$

$$d(I_i^n x, p) \leq (\frac{1}{2} + l_{in})[d(x, p) + d(p, I_i^n x)]$$

Let $k_n = \max\{k_{in} : i \in J\}$ and $l_n = \max\{l_{in} : i \in J\}$. So, $\{k_n\}, \{l_n\} \in [0, \infty)$ with $\lim_{n \to \infty} k_n = \lim_{n \to \infty} l_n = 0$. It follows that there exists $p \in F$ and $\{k_n\}, \{l_n\} \in [0, \infty)$ with $\lim_{n \to \infty} k_n = \lim_{n \to \infty} l_n = 0$, such that for all $x \in K$ and for each $i \in J$, the following holds

$$d(T_i^n x, p) \leq (\frac{1}{2} + k_n)[d(I_i^n x, p) + d(p, T_i^n x)]$$

$$d(I_i^n x, p) \leq (\frac{1}{2} + l_n)[d(x, p) + d(p, I_i^n x)]$$

Lemma B.2 1

Let (X, d, W) be a convex metric space with convex structure W, $\{T_i : i \in J\} : X \mapsto X$ be a finite family of I_i-asymptotically quasi-$(\delta, 1-\delta)$-nonexpansive mappings and $\{I_i : i \in J\} : X \mapsto X$ be a finite family of asymptotically quasi-$(\delta, 1-\delta)$ nonexpansive mapping with $F \neq \emptyset$. Suppose that $\sum_{n=1}^{\infty} k_n < \infty$, $\sum_{n=1}^{\infty} l_n < \infty$ and $\{x_n\}$ is as in Definition B.10 with $\{\gamma_n\}$ and $\{\widehat{\gamma}_n\}$ satisfying $\sum_{n=1}^{\infty} \gamma_n < \infty$ and $\sum_{n=1}^{\infty} \widehat{\gamma}_n < \infty$. If $\lim_{n \to \infty} d(x_n, F) = 0$ where $d(x, F) = \inf\{d(x, p) : p \in F\}$, then $\{x_n\}$ is a Cauchy sequence

Proof of Lemma B.2 1

Let $p \in F$. Since $\{u_n\}$ and $\{v_n\}$ are bounded sequences in X, there exists $M > 0$ such that

$$\max \left\{ \sup_{n \geq 1} d(u_n, p), \sup_{n \geq 1} d(v_n, p) \right\} \leq M$$

Now observe from Proposition B.1 and $\{x_n\}$ as in Definition B.10 , we have the following

$$
\begin{aligned}
d(y_n, p) &= d\left(W\left(x_n, T_i^n x_n, v_n; \widehat{\alpha}_n, \widehat{\beta}_n, \widehat{\gamma}_n \right), p \right) \\
&\leq \widehat{\alpha}_n d(x_n, p) + \widehat{\beta}_n d(T_i^n x_n, p) + \widehat{\gamma}_n d(v_n, p) \\
&\leq \widehat{\alpha}_n d(x_n, p) + \widehat{\beta}_n (\frac{1}{2} + k_n)[d(I_i^n x_n, p) + d(p, T_i^n x_n)] + \widehat{\gamma}_n M \\
&\leq \widehat{\alpha}_n d(x_n, p) + 2\widehat{\beta}_n (\frac{1}{2} + k_n) d(I_i^n x_n, p) + \widehat{\gamma}_n M \\
&\leq \widehat{\alpha}_n d(x_n, p) + 2\widehat{\beta}_n (\frac{1}{2} + k_n)(\frac{1}{2} + l_n)[d(x_n, p) + d(p, I_i^n x_n)] + \widehat{\gamma}_n M \\
&\leq \widehat{\alpha}_n d(x_n, p) + 4\widehat{\beta}_n (\frac{1}{2} + k_n)(\frac{1}{2} + l_n) d(x_n, p) + \widehat{\gamma}_n M \\
&\leq \widehat{\alpha}_n d(x_n, p) + 4\widehat{\beta}_n (1 + k_n)(1 + l_n) d(x_n, p) + \widehat{\gamma}_n M \\
&= \left[\widehat{\alpha}_n + 4\widehat{\beta}_n \{ 1 + l_n + k_n + k_n l_n \} \right] d(x_n, p) + \widehat{\gamma}_n M \\
&\leq 4\left[\widehat{\alpha}_n + \widehat{\beta}_n \{ 1 + l_n + k_n + k_n l_n \} \right] d(x_n, p) + \widehat{\gamma}_n M \\
&\leq 4\left[1 + \widehat{\beta}_n \{ l_n + k_n + k_n l_n \} \right] d(x_n, p) + \widehat{\gamma}_n M
\end{aligned}
$$

On the other hand

$$
\begin{aligned}
d(x_{n+1}, p) &= d\left(W\left(x_n, I_i^n y_n, u_n; \alpha_n, \beta_n, \gamma_n \right), p \right) \\
&\leq \alpha_n d(x_n, p) + \beta_n d(I_i^n y_n, p) + \gamma_n d(u_n, p) \\
&\leq \alpha_n d(\dot{x}_n, p) + \beta_n (\frac{1}{2} + l_n)[d(y_n, p) + d(p, I_i^n y_n)] + \gamma_n M \\
&\leq \alpha_n d(x_n, p) + 2\beta_n (\frac{1}{2} + l_n) d(y_n, p) + \gamma_n M \\
&\leq \alpha_n d(x_n, p) + 2\beta_n (1 + l_n) d(y_n, p) + \gamma_n M
\end{aligned}
$$

From the two chain of inequalities immediately above, we deduce the following

$$
\begin{aligned}
d(x_{n+1}, p) &\leq \alpha_n d(x_n, p) + 2\beta_n (1 + l_n) d(y_n, p) + \gamma_n M \\
&\leq \alpha_n d(x_n, p) + 8\beta_n (1 + l_n) \left[1 + \widehat{\beta}_n \{ l_n + k_n + k_n l_n \} \right] d(x_n, p) \\
&\quad + 2\beta_n (1 + l_n) \widehat{\gamma}_n M + \gamma_n M \\
&\leq \alpha_n d(x_n, p) + 8\beta_n (1 + l_n) d(x_n, p) + 8\beta_n (1 + l_n) \widehat{\beta}_n \{ l_n + k_n + k_n l_n \} d(x_n, p) \\
&\quad + (2\beta_n (1 + l_n) \widehat{\gamma}_n + \gamma_n) M \\
&\leq \left[1 + 8\beta_n l_n + 8\beta_n \widehat{\beta}_n (1 + l_n) \{ l_n + k_n + k_n l_n \} \right] d(x_n, p) + (2\beta_n (1 + l_n) \widehat{\gamma}_n + \gamma_n) M
\end{aligned}
$$

Put $j_n := 8\beta_n l_n + 8\beta_n \widehat{\beta}_n (1 + l_n) \{ l_n + k_n + k_n l_n \}$ and $z_n := (2\beta_n (1 + l_n) \widehat{\gamma}_n + \gamma_n) M$ with $\sum_{n=1}^{\infty} j_n < \infty$ and $\sum_{n=1}^{\infty} z_n < \infty$, then from the chain of inequalities immediately above we obtain

$$d(x_{n+1}, p) \leq (1 + j_n) d(x_n, p) + z_n$$

Proof of Lemma B.2 Continued 2

Hence

$$d(x_{n+1}, F) \leq (1 + j_n)d(x_n, F) + z_n$$

Now applying Lemma B.12 to the above inequality, it follows that $\lim_{n\to\infty} d(x_n, F)$ exists. Now we show that $\{x_n\}$ is Cauchy. Since $\sum_{n=1}^{\infty} j_n < \infty$, $1 + x \leq e^x$ for all $x \geq 0$, then it follows from $d(x_{n+1}, p) \leq (1 + j_n)d(x_n, p) + z_n$ that

$$d(x_{n+1}, p) \leq \exp\{j_n\}d(x_n, p) + z_n$$

Hence for any positive integers n, m it follows from the above inequality that we have the following

$$
\begin{aligned}
d(x_{n+m}, p) &\leq \exp\{j_{n+m-1}\}d(x_{n+m-1}, p) + z_{n+m-1} \\
&\leq \exp\{j_{n+m-1}\}[\exp\{j_{n+m-2}\}d(x_{n+m-2}, p) + z_{n+m-2}] + z_{n+m-1} \\
&= \exp\{j_{n+m-1}\}\exp\{j_{n+m-2}\}d(x_{n+m-2}, p) + \exp\{j_{n+m-1}\}z_{n+m-2} + z_{n+m-1} \\
&\leq \cdots \\
&\leq \exp\left\{\sum_{i=n}^{n+m-1} j_i\right\}d(x_n, p) + \exp\left\{\sum_{i=n}^{n+m-1} j_i\right\}\sum_{i=n}^{n+m-1} z_i \\
&\leq Hd(x_n, p) + H\sum_{i=n}^{n+m-1} z_i
\end{aligned}
$$

where $H = \exp\left\{\sum_{i=n}^{n+m-1} j_i\right\} < \infty$ Since $\lim_{n\to\infty} d(x_n, F) = 0$ and $\sum_{n=1}^{\infty} z_n < \infty$, then for any given $\epsilon > 0$ there exists a positive integer n_0 such that $d(x_n, F) < \frac{\epsilon}{4(H+1)}$ and $\sum_{n=1}^{\infty} z_n < \frac{\epsilon}{2H}$, for all $n \geq n_0$. It follows that there exists $p_1 \in F$ such that for all $n \geq n_0$

$$d(x_n, p_1) < \frac{\epsilon}{2(H+1)}$$

Consequently, for any $n \geq n_0$ and for all $m \geq 1$ we have

$$
\begin{aligned}
d(x_{n+m}, x_n) &\leq d(x_{n+m}, p_1) + d(x_n, p_1) \\
&\leq (1 + H)d(x_n, p_1) + H\sum_{n=1}^{\infty} z_n \\
&\leq \frac{\epsilon}{2(H+1)}(H+1) + H\frac{\epsilon}{2H} \\
&= \epsilon
\end{aligned}
$$

It follows that $\{x_n\}$ is a Cauchy sequence in X

Remark B.3 1

[I. Yıldırım and S.H. Khan, Convergence theorems for common fixed points of asymptotically quasi-nonexpansive mappings in convex metric spaces, Appl. Math. Comput. 218 (2012) 4860–4866] Lemma B.12(b) holds under the hypothesis $\limsup_{n\to\infty} a_n = 0$ as well. Therefore condition (b) in Lemma B.12 can be reformulated as: if either $\liminf_{n\to\infty} a_n = 0$ or $\limsup_{n\to\infty} a_n = 0$, then $\lim_{n\to\infty} a_n = 0$

Theorem B.4 1

Let (X, d, W) be a convex metric space with convex structure W, $\{T_i : i \in J\} : X \mapsto X$ be a finite family of I_i-asymptotically quasi-$(\delta, 1 - \delta)$-nonexpansive mappings and $\{I_i : i \in J\} : X \mapsto X$ be a finite family of asymptotically quasi-$(\delta, 1 - \delta)$ nonexpansive mapping with $F \neq \emptyset$. Suppose that $\sum_{n=1}^{\infty} k_n < \infty$, $\sum_{n=1}^{\infty} l_n < \infty$ and $\{x_n\}$ is as in Definition B.10 with $\{\gamma_n\}$ and $\{\widehat{\gamma}_n\}$ satisfying $\sum_{n=1}^{\infty} \gamma_n < \infty$ and $\sum_{n=1}^{\infty} \widehat{\gamma}_n < \infty$. Then

(a) $\liminf_{n \to \infty} d(x_n, F) = \limsup_{n \to \infty} d(x_n, F) = 0$ if $\{x_n\}$ converges to a unique fixed point in F

(b) $\{x_n\}$ converges to a unique fixed point in F if X is complete and either $\liminf_{n \to \infty} d(x_n, F) = 0$ or $\limsup_{n \to \infty} d(x_n, F) = 0$

Proof of Theorem B.4 1

Let $p \in F$. Since $\{x_n\}$ converges to p, $\lim_{n \to \infty} d(x_n, p) = 0$. So, for a given $\epsilon > 0$, there exists $n_0 \in \mathbb{N}$ such that $d(x_n, p) < \epsilon$ for all $n \geq n_0$. Taking infimum over $p \in F$, we have for all $n \geq n_0$, $d(x_n, F) < \epsilon$, which implies that $\lim_{n \to \infty} d(x_n, F) = 0$. It follows that $\lim_{n \to \infty} \inf d(x_n, F) = \lim_{n \to \infty} \sup d(x_n, F) = 0$, and (a) is proved.

To see (b), suppose X is complete and $\lim_{n \to \infty} \inf d(x_n, F) = 0$ or $\lim_{n \to \infty} \sup d(x_n, F) = 0$. Then from Lemma B.12(b) and Remark B.3, it follows that $\lim_{n \to \infty} d(x_n, F) = 0$. From the completeness of X and Lemma B.2, we get that $\lim_{n \to \infty} x_n$ exists. Put $\lim_{n \to \infty} x_n = q \in X$, we prove that $q \in F$. For a given $\epsilon_1 > 0$, there exists a constant n_1 such that for all $n \geq n_1$, we have, $d(x_n, q) < \frac{\epsilon_1}{2(2+2l_1)}$ and $d(x_n, F) < \frac{\epsilon_1}{2(4+6l_1)}$. In particular, there exists $s \in F$ and a constant $n_2 \geq n_1$ such that $d(x_{n_2}, s) < \frac{\epsilon_1}{2(4+6l_1)}$. Now for any I_i, $i \in J$, observe we have the following

$$
\begin{aligned}
d(I_i q, q) &\leq d(I_i q, s) + d(s, I_i x_{n_2}) + d(I_i x_{n_2}, s) + d(s, x_{n_2}) + d(x_{n_2}, q) \\
&= d(I_i q, s) + 2d(s, I_i x_{n_2}) + d(s, x_{n_2}) + d(x_{n_2}, q) \\
&\leq (\frac{1}{2} + l_1)[d(q, s) + d(q, I_i s)] + 2(\frac{1}{2} + l_1)[d(s, x_{n_2}) + d(s, I_i x_{n_2})] \\
&\quad + d(s, x_{n_2}) + d(x_{n_2}, q) \\
&\leq 2(\frac{1}{2} + l_1)d(q, s) + 4(\frac{1}{2} + l_1)d(s, x_{n_2}) + d(s, x_{n_2}) + d(x_{n_2}, q) \\
&= (1 + 2l_1)d(q, s) + (3 + 4l_1)d(s, x_{n_2}) + d(x_{n_2}, q) \\
&\leq (1 + 2l_1)d(q, x_{n_2}) + (1 + 2l_1)d(x_{n_2}, s) + (3 + 4l_1)d(s, x_{n_2}) + d(x_{n_2}, q) \\
&= (2 + 2l_1)d(q, x_{n_2}) + (4 + 6l_1)d(x_{n_2}, s) \\
&\leq (2 + 2l_1)\frac{\epsilon_1}{2(2 + 2l_1)} + (4 + 6l_1)\frac{\epsilon_1}{2(4 + 6l_1)} \\
&= \epsilon_1
\end{aligned}
$$

Since ϵ_1 is arbitrary, we have $d(I_i q, q) = 0$ for all $i \in J$, that is, $I_i q = q$. This implies that $q \in \bigcap_{i=1}^{k} F(I_i)$. Similarly, $q \in \bigcap_{i=1}^{k} F(T_i)$. Therefore $q \in F$

2.4   Exercises

Recall from [Clement Boateng Ampadu, A Strong Ciric Almost Contraction Mapping Theorem in Metric Spaces with Unique Fixed Point, Applied Mathematics(AM), Initially Accepted but Withdrawn] that if (X, d) is a metric space, then a map $T : X \mapsto X$ is called a strong Ciric $(\alpha, 1 - \alpha)$-weak contraction if there exists $\alpha \in (0, 1)$ such that for all $x, y \in X$ the following holds

$$d(Tx, Ty) \leq \alpha M_1(x, y) + (1 - \alpha)d(y, Tx)$$

where

$$M_1(x,y) = \max\{d(x,y), d(x,Tx), d(y,Ty), \frac{1}{2}[d(x,Ty) + d(y,Tx)]\}$$

Observe

$$d(Tx,Ty) \leq \alpha M_1(x,y) + (1-\alpha)d(y,Tx)$$

where

$$M_1(x,y) = \max\{d(x,y), d(x,Tx), d(y,Ty), \frac{1}{2}[d(x,Ty) + d(y,Tx)]\}$$

is equivalent to

$$d(Tx,Ty) \leq (\alpha + 1 - \alpha)\max\{M_1(x,y), d(y,Tx)\}$$

where

$$M_1(x,y) = \max\{d(x,y), d(x,Tx), d(y,Ty), \frac{1}{2}[d(x,Ty) + d(y,Tx)]\}$$

Now we will say that T is I-asymptotically quasi-$(\alpha, 1-\alpha)$-nonexpansive if $F(T) \cap F(I) \neq \emptyset$ and there exists a sequence $\{v_n\} \subset [0, \infty)$ with $\lim_{n\to\infty} v_n = 0$ such that

$$d(T^n x, p) \leq (\frac{1}{2} + v_n)[M_3(x,p) + d(p, T^n x)]$$

for all $x \in X$, $p \in F(T) \cap F(I)$ and $n \geq 1$, where

$$M_3(x,p) = \max\{d(I^n x, p), d(I^n x, T^n x), 0, \frac{1}{2}[d(I^n x, p) + d(p, T^n x)]\}$$

Also we say T is asymptotically quasi-$(\alpha, 1-\alpha)$-nonexpansive if $F(T) \neq \emptyset$ and there there exists $u_n \in [0, \infty)$ for all $n \in \mathbb{N}$ with $\lim_{n\to\infty} u_n = 0$ such that

$$d(T^n x, p) \leq (\frac{1}{2} + u_n)[M_4(x,p) + d(p, T^n x)]$$

for all $x \in X$, $p \in F(T)$, and $n \in \mathbb{N}$, where

$$M_4(x,p) = \max\{d(x,p), d(x, T^n x), 0, \frac{1}{2}[d(x,p) + d(p, T^n x)]\}$$

Exercise B.1 1

Let (X,d) be a convex metric space with convex structure W, $\{T_i : i \in J\} : X \mapsto X$ be a finite family of I_i-asymptotically quasi-$(\alpha, 1-\alpha)$-nonexpansive mappings and $\{I_i : i \in J\} : X \mapsto X$ be a finite family of asymptotically quasi-$(\alpha, 1-\alpha)$ nonexpansive mappings with $F := \left(\bigcap_{i=1}^r F(T_i)\right) \cap \left(\bigcap_{i=1}^r F(I_i)\right) \neq \emptyset$. Then there exists a point $p \in F$ and sequences $\{k_n\}, \{l_n\} \in [0, \infty)$ with $\lim_{n\to\infty} k_n = \lim_{n\to\infty} l_n = 0$ such that for all $x \in K$ and for each $i \in I$, the following inequalities hold

(a)

$$d(T_i^n x, p) \leq (\frac{1}{2} + k_n)[M_5(x,p) + d(p, T_i^n x)]$$

where

$$M_5(x,p) = \max\{d(I_i^n x, p), d(I_i^n x, T_i^n x), 0, \frac{1}{2}[d(I_i^n x, p) + d(p, T_i^n x)]\}$$

(b)

$$d(I_i^n x, p) \leq (\frac{1}{2} + l_n)[M_6(x,p) + d(p, I_i^n x)]$$

where

$$M_6(x,p) = \max\{d(x,p), d(x, I_i^n x), 0, \frac{1}{2}[d(x,p) + d(p, I_i^n x)]\}$$

Exercise B.2 1

Let (X, d, W) be a convex metric space with convex structure W, $\{T_i : i \in J\} : X \mapsto X$ be a finite family of I_i-asymptotically quasi-$(\alpha, 1 - \alpha)$-nonexpansive mappings and $\{I_i : i \in J\} : X \mapsto X$ be a finite family of asymptotically quasi-$(\alpha, 1 - \alpha)$ nonexpansive mapping with $F \neq \emptyset$. Suppose that $\sum_{n=1}^{\infty} k_n < \infty$, $\sum_{n=1}^{\infty} l_n < \infty$ and $\{x_n\}$ is as in Definition B.10 with $\{\gamma_n\}$ and $\{\widehat{\gamma}_n\}$ satisfying $\sum_{n=1}^{\infty} \gamma_n < \infty$ and $\sum_{n=1}^{\infty} \widehat{\gamma}_n < \infty$. If $\lim_{n\to\infty} d(x_n, F) = 0$ where $d(x, F) = \inf\{d(x, p) : p \in F\}$, then $\{x_n\}$ is a Cauchy sequence

---

Exercise B.3 1

Let (X, d, W) be a convex metric space with convex structure W, $\{T_i : i \in J\} : X \mapsto X$ be a finite family of I_i-asymptotically quasi-$(\alpha, 1 - \alpha)$-nonexpansive mappings and $\{I_i : i \in J\} : X \mapsto X$ be a finite family of asymptotically quasi-$(\alpha, 1 - \alpha)$ nonexpansive mapping with $F \neq \emptyset$. Suppose that $\sum_{n=1}^{\infty} k_n < \infty$, $\sum_{n=1}^{\infty} l_n < \infty$ and $\{x_n\}$ is as in Definition B.10 with $\{\gamma_n\}$ and $\{\widehat{\gamma}_n\}$ satisfying $\sum_{n=1}^{\infty} \gamma_n < \infty$ and $\sum_{n=1}^{\infty} \widehat{\gamma}_n < \infty$. Then

(a) $\liminf_{n\to\infty} d(x_n, F) = \limsup_{n\to\infty} d(x_n, F) = 0$ if $\{x_n\}$ converges to a unique fixed point in F

(b) $\{x_n\}$ converges to a unique fixed point in F if X is complete and either $\liminf_{n\to\infty} d(x_n, F) = 0$ or $\limsup_{n\to\infty} d(x_n, F) = 0$

2.5 References

(1) Clement Boateng Ampadu, Unique Fixed Point Theorem for Weakly $(\delta, 1 - \delta)$-weak Contractive Mappings, Unpublished. Available online:
https://drive.google.com/file/d/0BwtkpMtWoUlEdXVNQVBGUVRWbEU/view

(2) Clement Boateng Ampadu, An Almost Contraction Mapping Theorem in Metric Spaces with Unique Fixed Point, Submitted. Available online:
https://drive.google.com/file/d/0BwtkpMtWoUlEY25OZW1HUEdGcU0/view

(3) Birol Gunduz, Fixed Points of a Finite Family of I-Asymptotically Quasi-Nonexpansive Mappings in a Convex Metric Space, Filomat 31:7 (2017), 2175 2182

(4) S.Temir, On the convergence theorems of implicit iteration process for a finite family of I-asymptotically nonexpansive mappings, J. Comput. Appl. Math. 225 (2009) 398–405

(5) S. Temir, O. Gul, Convergence theorem for I-asymptotically quasi-nonexpansive mapping in Hilbert space, J. Math. Anal. Appl.329 (2007) 759–765

(6) W. Takahashi, A convexity in metric space and nonexpansive mappings, Kodai. Math. Sem. Rep. 22 (1970) 142-149

(7) L. Qihou, Iterative sequence for asymptotically quasi-nonexpansive mappings with errors member, J. Math. Anal. Appl. 259 (2001) 18–24

(8) I. Yıldırım and S.H. Khan, Convergence theorems for common fixed points of asymptotically quasi-nonexpansive mappings in convex metric spaces, Appl. Math. Comput. 218 (2012) 4860–4866

(9) Clement Boateng Ampadu, A Strong Ciric Almost Contraction Mapping Theorem in Metric Spaces with Unique Fixed Point, Applied Mathematics(AM), Initially Accepted but Withdrawn. Available online:
https://drive.google.com/file/d/0BwtkpMtWoUlEZHhpdExKVGdMOTA/view

Chapter 3

Strong Convergence Theorems for a Finite Family of Nonself I-Asymptotically and Asymptotically $(\delta, 1 - \delta)$ Nonexpansive Mappings

3.1 Brief Summary

Abstract C.1 1

We introduce non-self asymptotically $I - (\delta, 1 - \delta)$-nonexpansive mappings and non-self asymptotically $(\delta, 1 - \delta)$ nonexpansive mappings and use the explicit iterative sequence scheme introduced in [Birol Gündüz and Sezgin Akbulut, On Weak and Strong Convergence Theorems for a Finite Family of Nonself I-asymptotically Nonexpansive Mappings, Mathematica Moravica Vol. 19-2 (2015), 49–64] to obtain some strong convergence theorems of this iterative scheme for a family of non-self asymptotically $I - (\delta, 1 - \delta)$-nonexpansive mappings $\{T_i\}_i^N$ and non-self asymptotically $(\delta, 1 - \delta)$ nonexpansive mappings $\{I_i\}_i^N$

3.2 Preliminaries

Definition C.1 1

[Birol Gündüz and Sezgin Akbulut, On Weak and Strong Convergence Theorems for a Finite Family of Nonself I-asymptotically Nonexpansive Mappings, Mathematica Moravica Vol. 19-2 (2015), 49–64] Let X be a real normed linear space, K be a nonempty subset of X, and T be a self-mapping of K.

(a) T is said to be nonexpansive if $\|Tx - Ty\| \leq \|x - y\|$ holds for all $x, y \in K$

(b) T is said to be asymptotically nonexpansive if there exists a sequence $\{k_n\} \in [0, \infty)$, with $\lim_{n \to \infty} k_n = 0$, such that $\|T^n x - T^n y\| \leq (1 + k_n)\|x - y\|$ holds for all $x, y \in K$ and $n \geq 1$

(c) T is called uniformly L-Lipschitzian if there exists a constant $L > 0$ such that

$$\|T^n x - T^n y\| \leq L\|x - y\|$$

for all $x, y \in K$ and positive integer $n \geq 1$

Recall the concept of $(\delta, 1 - \delta)$-weak contraction was introduced in [Clement Boateng Ampadu, An Almost Contraction Mapping Theorem in Metric Spaces with Unique Fixed Point, Submitted] as follows. A map $T : X \mapsto X$ is called a $(\delta, 1 - \delta)$-weak contraction if there exists $\delta \in (0, 1)$ such that $d(Tx, Ty) \leq \delta d(x, y) + (1 - \delta)d(y, Tx)$ for all $x, y \in X$. Observe that

$$d(Tx, Ty) \leq \delta d(x, y) + (1 - \delta)d(y, Tx)$$

is equivalent to

$$d(Tx, Ty) \leq (\delta + 1 - \delta) \max\{d(x, y), d(y, Tx)\}$$

This observation and the above definition motivates the following

Definition C.2 1

Let X be a real normed linear space, K be a nonempty subset of X, and $T : K \mapsto K$ be a $(\delta, 1 - \delta)$-weak contraction

(a) T will be called $(\delta, 1 - \delta)$-nonexpansive if $\|Tx - Ty\| \leq \frac{1}{2}\big[\|x - y\| + \|y - Tx\|\big]$ holds for all $x, y \in K$

(b) T will be called asymptotically $(\delta, 1 - \delta)$-nonexpansive if there exists a sequence $\{k_n\} \in [0, \infty)$, with $\lim_{n \to \infty} k_n = 0$, such that $\|T^n x - T^n y\| \leq (\frac{1}{2} + k_n)\big[\|x - y\| + \|y - T^n x\|\big]$ holds for all $x, y \in K$ and $n \geq 1$

(c) T will be called uniformly L-Lipschitzian if there exists a constant $L > 0$ such that $\|T^n x - T^n y\| \leq L\big[\|x - y\| + \|y - T^n x\|\big]$ holds for all $x, y \in K$ and $n \geq 1$

Note that the concept of asymptotically non-expansive self-mapping was proposed by Goebel and Kirk [K. Goebel, W.A. Kirk, A fixed point theorem for asymptotically nonexpansive mappings, Proc. Amer. Math. Soc., 35 (1972), 171–174], in which it is proved that if X is uniformly convex, and K is a bounded closed and convex subset of X, then every asymptotically nonexpansive self mapping has a fixed point. Clearly, we have the following

Question C.3 1

Suppose X is uniformly convex, and K is a bounded closed and convex subset of X, does every asymptotically $(\delta, 1 - \delta)$-nonexpansive self mapping has a fixed point?

Definition C.4 1

A subset K of X is said to be a retract of X if there exists a continuous map $P : X \mapsto K$ such that $Px = x$, for all $x \in K$

Note that every closed convex subset of a uniformly convex Banach space is a retract.

Definition C.5 1

A map $P : X \mapsto K$ is called a retraction if $P^2 = P$

Note that if a map P is a retraction, then $Py = y$ for all y in the range of P. The concept of nonself asymptotically nonexpansive mapping was proposed in [C.E. Chidume, E.U. Ofoedu, H. Zegeye, Strong and weak convergence theorems for asymptotically nonexpansive mappings, J. Math. Anal. Appl., 280 (2003), 364-374] as further generalization of the concept of asymptotically nonexpansive self-mapping. In particular they introduced the following

---

Definition C.6 1

[C.E. Chidume, E.U. Ofoedu, H. Zegeye, Strong and weak convergence theorems for asymptotically nonexpansive mappings, J. Math. Anal. Appl., 280 (2003), 364-374] Let K be a nonempty subset of a real normed space X and $P : X \mapsto K$ be a nonexpansive retraction of X onto K

(a) $T : K \mapsto X$ is said to be asymptotically nonexpansive if there exists a sequence $\{k_n\} \in [0, \infty)$, with $\lim_{n \to \infty} k_n = 0$, such that $\|T(PT)^{n-1}x - T(PT)^{n-1}y\| \leq (1 + k_n)\|x - y\|$ holds for all $x, y \in K$ and any positive integer n

(b) $T : K \mapsto X$ is called uniformly L-Lipschitzian if there exists a constant $L > 0$ such that $\|T(PT)^{n-1}x - T(PT)^{n-1}y\| \leq L\|x - y\|$ for all $x, y \in K$

Now as a further generalization of the concept of asymptotically $(\delta, 1 - \delta)$-nonexpansive self-mapping we introduce the following

Definition C.7 1

Let K be a nonempty subset of a real normed space X and $P : X \mapsto K$ be a $(\delta, 1 - \delta)$-nonexpansive retraction of X onto K

(a) $T : K \mapsto X$ will be called asymptotically $(\delta, 1 - \delta)$ nonexpansive if there exists a sequence $\{k_n\} \in [0, \infty)$, with $\lim_{n \to \infty} k_n = 0$, such that

$$\|T(PT)^{n-1}x - T(PT)^{n-1}y\| \leq (\frac{1}{2} + k_n)\big[\|x - y\| + \|y - T(PT)^{n-1}x\|\big]$$

holds for all $x, y \in K$ and any positive integer n

(b) $T : K \mapsto X$ will be called uniformly L-Lipschitzian if there exists a constant $L > 0$ such that $\|T(PT)^{n-1}x - T(PT)^{n-1}y\| \leq L\big[\|x - y\| + \|y - T(PT)^{n-1}x\|\big]$ for all $x, y \in K$

Note that if P is the identity in Definition C.7(a), then we recover Definition C.2(b). Similarly, if P is the identity in Definition C.7(b), then we recover Definition C.2(c)

Definition C.8 1

[S. Temir, On the convergence theorems of implicit iteration process for a finite family of I-asymptotically nonexpansive mappings, J. Comput. Appl. Math. 225 (2009),398–405; S. Temir, O. Gul, Convergence theorem for I-asymptotically quasi-nonexpansive mapping in Hilbert space, J. Math. Anal. Appl. 329 (2007), 759–765] Let K be a nonempty subset of a real normed linear space E. For self mappings $T, I : K \mapsto K$

(a) T is called asymptotically I-nonexpansive on K if there exists a sequence $\{v_n\} \in [0, \infty)$, with $\lim_{n \to \infty} v_n = 0$, such that $\|T^n x - T^n y\| \leq (1 + v_n)\|I^n x - I^n y\|$ holds for all $x, y \in K$ and $n \geq 1$

(b) T is called I-uniformly Lipschitzian if there exists $\Gamma > 0$ such that

$$\|T^n x - T^n y\| \leq \Gamma \|I^n x - I^n y\|$$

holds for all $x, y \in K$ and $n \geq 1$

Now we introduce the following

> **Definition C.9 1**
>
> Let K be a nonempty subset of a real normed linear space E. For self mappings $T, I : K \mapsto K$, where $T : K \mapsto K$ is a $(\delta, 1 - \delta)$-weak contraction
>
> (a) T will be called asymptotically I-$(\delta, 1-\delta)$-nonexpansive on K if there exists a sequence $\{v_n\} \in [0, \infty)$, with $\lim_{n \to \infty} v_n = 0$, such that
>
> $$\|T^n x - T^n y\| \leq (\frac{1}{2} + v_n)\big[\|I^n x - I^n y\| + \|I^n y - T^n x\|\big]$$
>
> holds for all $x, y \in K$ and $n \geq 1$
>
> (b) T will be called I-uniformly Lipschitzian if there exists $\Gamma > 0$ such that
>
> $$\|T^n x - T^n y\| \leq \Gamma\big[\|I^n x - I^n y\| + \|I^n y - T^n x\|\big]$$
>
> holds for all $x, y \in K$ and $n \geq 1$

The class of I-asymptotically nonexpansive nonself mappings, it seems, appeared first in [L. Yang, X. Xie, Weak and strong convergence theorems for a finite family of I-asymptotically nonexpansive mappings, Appl. Math. Comput. 216 (2010), 1057–1064]. In particular they introduced the following

> **Definition C.10 1**
>
> [L. Yang, X. Xie, Weak and strong convergence theorems for a finite family of I-asymptotically nonexpansive mappings, Appl. Math. Comput. 216 (2010), 1057–1064]. Let K be a nonempty subset of a real normed space X and $P : X \mapsto K$ be a nonexpansive retraction of X onto K. Let $T, I : K \mapsto X$ be two mappings
>
> (a) $T : K \mapsto X$ is said to be I-asymptotically nonexpansive if there exists a sequence $\{u_n\} \in [0, \infty)$, with $\lim_{n \to \infty} u_n = 0$, such that
>
> $$\|T(PT)^{n-1}x - T(PT)^{n-1}y\| \leq (1 + u_n)\|I(PI)^{n-1}x - I(PI)^{n-1}y\|$$
>
> holds for all $x, y \in K$ and any positive integer n
>
> (b) $T : K \mapsto X$ is called uniformly Γ-Lipschitzian if there exists a constant $\Gamma > 0$ such that
> $$\|T(PT)^{n-1}x - T(PT)^{n-1}y\| \leq \Gamma\|I(PI)^{n-1}x - I(PI)^{n-1}y\|$$
> for all $x, y \in K$

Now we introduce the following

Definition C.11 1

Let K be a nonempty subset of a real normed space X and $P : X \mapsto K$ be a $(\delta, 1 - \delta)$-nonexpansive retraction of X onto K. Let $T, I : K \mapsto X$ be two mappings.

(a) $T : K \mapsto X$ will be called I-asymptotically $(\delta, 1 - \delta)$ nonexpansive if there exists a sequence $\{u_n\} \in [0, \infty)$, with $\lim_{n \to \infty} u_n = 0$, such that

$$\|T(PT)^{n-1}x - T(PT)^{n-1}y\| \leq (\frac{1}{2} + u_n) \big[\|I(PI)^{n-1}x - I(PI)^{n-1}y\| $$
$$+ \|I(PI)^{n-1}y - T(PT)^{n-1}x\| \big]$$

holds for all $x, y \in K$ and any positive integer n

(b) $T : K \mapsto X$ will be called uniformly Γ-Lipschitzian if there exists a constant $\Gamma > 0$ such that

$$\|T(PT)^{n-1}x - T(PT)^{n-1}y\| \leq \Gamma \big[\|I(PI)^{n-1}x - I(PI)^{n-1}y\| + \|I(PI)^{n-1}y - T(PT)^{n-1}x\| \big]$$

for all $x, y \in K$

The following iterative sequence was introduced in [Birol Gündüz and Sezgin Akbulut, On Weak and Strong Convergence Theorems for a Finite Family of Nonself I-asymptotically Nonexpansive Mappings, Mathematica Moravica Vol. 19-2 (2015), 49–64]: Let K be a nonempty subset of a Banach space X. Let $\{T_i\}_i^N : K \mapsto X$ be N nonself I_i-asymptotically nonexpansive mappings and $\{I_i\}_i^N : K \mapsto X$ be N nonself asymptotically nonexpansive mappings. Let $\{\alpha_n\}$ and $\{\beta_n\}$ be two real sequences in $[0, 1]$. Then the sequence $\{x_n\}$ is generated as follows for all $n \geq 1$

$$x_{n+1} = P\bigg((1 - \alpha_n)T_i(PT_i)^{n-1}x_n + \alpha_n I_i(PI_i)^{n-1}y_n \bigg)$$

$$y_n = P\bigg((1 - \beta_n)x_n + \beta_n T_i(PT_i)^{n-1}x_n \bigg)$$

where $n = (k - 1)N + i$, $i = i(n) \in J := \{1, 2, \cdots, N\}$ is a positive integer and $k(n) \to \infty$ as $n \to \infty$. Note that the sequence $\{x_n\}$ can also be generated as follows for all $n \geq 1$

$$x_{n+1} = P\bigg((1 - \alpha_n)T_i(PT_i)^{k(n)-1}x_n + \alpha_n I_i(PI_i)^{k(n)-1}y_n \bigg)$$

$$y_n = P\bigg((1 - \beta_n)x_n + \beta_n T_i(PT_i)^{k(n)-1}x_n \bigg)$$

In this paper we obtain some strong convergence theorems of the above iterative scheme for a family of non-self asymptotically $I - (\delta, 1 - \delta)$-nonexpansive mappings $\{T_i\}_i^N$ and non-self asymptotically $(\delta, 1 - \delta)$ nonexpansive mappings $\{I_i\}_i^N$. The remaining concepts and results presented below will be useful in the sequel.

Definition C.12 1

[Birol Gündüz and Sezgin Akbulut, On Weak and Strong Convergence Theorems for a Finite Family of Nonself I-asymptotically Nonexpansive Mappings, Mathematica Moravica Vol. 19-2 (2015), 49–64] A Banach space X is said to satisfy Opial's condition if, for any sequence $\{x_n\}$ in X, $\{x_n\}$ converges weakly to x implies that

$$\limsup_{n \to \infty} \|x_n - x\| < \limsup_{n \to \infty} \|x_n - y\|$$

for all $y \in X$ with $y \neq x$

Definition C.13 1

[M.O. Osilike, A. Udomene, Demiclosedness principle and convergence theorems for strictly pseudocontractive mappings of Browder–Petryshyn type, J. Math. Anal. Appl. 256 (2001), 431–445] A Banach space X is said to have a Frechet differentiable norm if for all $x \in S_X = \{x \in X : \|x\| = 1\}$

$$\lim_{t \to 0} \frac{\|x + ty\| - \|x\|}{t}$$

exists and is attained uniformly in $y \in S_X$

Definition C.14 1

[Birol Gündüz and Sezgin Akbulut, On Weak and Strong Convergence Theorems for a Finite Family of Nonself I-asymptotically Nonexpansive Mappings, Mathematica Moravica Vol. 19-2 (2015), 49–64] A mapping T with domain $D(T)$ and range $R(T)$ in X is said to be demiclosed at p if whenever $\{x_n\}$ is a sequence in $D(T)$ such that $x_n \to x^* \in D(T)$ and $\{Tx_n\}$ converges weakly to p, then $Tx^* = p$

Definition C.15 1

[Birol Gündüz and Sezgin Akbulut, On Weak and Strong Convergence Theorems for a Finite Family of Nonself I-asymptotically Nonexpansive Mappings, Mathematica Moravica Vol. 19-2 (2015), 49–64] A mapping $T : K \mapsto K$ is said to be semicompact if, for any bounded sequence $\{x_n\}$ in K such that $\lim_{n \to \infty} \|x_n - Tx_n\| = 0$, there exists a subsequence $\{x_{n_j}\}$ of $\{x_n\}$ such that $\{x_{n_j}\}$ converges strongly to some x^* in K

Definition C.16 1

[H.F. Senter, W.G. Dotson, Approximating fixed points of nonexpansive mappings, Proc. Amer. Math. Soc., 44 (1974), 375-380] The mapping $T : K \mapsto K$ with $F(T) \neq \emptyset$ is said to satisfy condition (A) if there is a nondecreasing function $f : [0, \infty) \mapsto [0, \infty)$ with $f(0) = 0$, $f(t) > 0$ for all $t \in (0, \infty)$ such that $\|x - Tx\| \geq f(d(x, F(T)))$ for all $n \geq 1$

Note that every continuous and semi-compact mapping must satisfy condition (A) [H.F. Senter, W.G. Dotson, Approximating fixed points of nonexpansive mappings, Proc. Amer. Math. Soc., 44 (1974), 375-380]

Definition C.17 1

[S.H. Khan, H. Fukharuddin, Weak and strong convergence of a scheme with errors for two nonexpansive mappings, Nonlinear Anal. 61 (2005), 1295–1301] Two mappings $T_1, T_2 : K \mapsto K$ are said to satisfy condition (A') if there is a nondecreasing function $f : [0, \infty) \mapsto [0, \infty)$ with $f(0) = 0$, $f(t) > 0$ for all $t \in (0, \infty)$ such that

$$\frac{1}{2}[\|x - T_1 x\| + \|x - T_2 x\|] \geq f(d(x, F))$$

for all $x \in K$, where $d(x, F) = \inf\{\|x - p\| : p \in F := F(T_1) \cap F(T_2)\}$

By modifying the above two definitions, the following was obtained

Definition C.18 1

[Birol Gündüz and Sezgin Akbulut, On Weak and Strong Convergence Theorems for a Finite Family of Nonself I-asymptotically Nonexpansive Mappings, Mathematica Moravica Vol. 19-2 (2015), 49–64] A family $\{T_i\}_i^N : K \mapsto X$ of N nonself I_i-asymptotically nonexpansive mappings and a family $\{I_i\}_i^N : K \mapsto X$ of N nonself asymptotically nonexpansive mappings with $F = \bigcap_{i=1}^N F(T_i) \cap F(I_i) \neq \emptyset$ are said to satisfy condition (B) if there is a nondecreasing function $f : [0, \infty) \mapsto [0, \infty)$ with $f(0) = 0$, $f(t) > 0$ for all $t \in (0, \infty)$ such that

$$\max_{1 \leq i \leq N} \left[\frac{1}{2} [\|x - T_i x\| + \|x - I_i x\|] \right] \geq f(d(x, F))$$

Finally we will need the following two Lemma's

Lemma C.19 1

[K.K. Tan and H.K. Xu, Approximating fixed points of nonexpansive mappings by the Ishikawa iteration process, Journal of Mathematical Analysis and Applications, vol. 178, no. 2 (1993), pp. 301–308] Let $\{a_n\}$, $\{b_n\}$, and $\{\delta_n\}$ be sequences of nonnegative real numbers satisfying the inequality

$$a_{n+1} \leq (1 + \delta_n) a_n + b_n$$

If $\sum_{n=1}^\infty b_n < \infty$ and $\sum_{n=1}^\infty \delta_n < \infty$, then $\lim_{n \to \infty} a_n$ exists

Lemma C.20 1

[J. Schu, Weak and strong convergence to fixed points of asymptotically nonexpansive mappings, Bull. Austral. Math. Soc. 43 (1991), 153-159] Suppose that X is a uniformly convex Banach space and $0 < p \leq t_n \leq q < 1$ for all $n \geq 1$. Also suppose that $\{x_n\}$ and $\{y_n\}$ are sequences in X such that $\lim_{n \to \infty} \sup \|x_n\| \leq r$, $\lim_{n \to \infty} \sup \|y_n\| \leq r$, and $\lim_{n \to \infty} \|t_n x_n + (1 - t) y_n\| = r$ holds for some $r \geq 0$. Then $\lim_{n \to \infty} \|x_n - y_n\| = 0$

3.3   Main Results

Lemma C.1 1

Let X be a real Banach space, K be a nonempty closed convex subset of X which is also a $(\delta, 1 - \delta)$ nonexpansive retract with retraction P. Let $\{T_i\}_i^N : K \mapsto X$ be N nonself I_i-asymptotically $(\delta, 1-\delta)$ nonexpansive mappings with sequences $\{l_n^{(i)}\} \subset [0, \infty)$ such that $\sum_{n=1}^\infty l_n^{(i)} < \infty$ and $\{I_i\}_i^N : K \mapsto X$ be N nonself asymptotically $(\delta, 1-\delta)$ nonexpansive mappings with sequences $\{k_n^{(i)}\} \subset [0, \infty)$ such that $\sum_{n=1}^\infty k_n^{(i)} < \infty$. Suppose that for any given $x_1 \in K$, the sequence $\{x_n\}$ is generated as follows for all $n \geq 1$

$$x_{n+1} = P\left((1 - \alpha_n) T_i (PT_i)^{n-1} x_n + \alpha_n I_i (PI_i)^{n-1} y_n \right)$$

$$y_n = P\left((1 - \beta_n) x_n + \beta_n T_i (PT_i)^{n-1} x_n \right)$$

and $F = \bigcap_{i=1}^N F(T_i) \cap F(I_i) \neq \emptyset$. Then

(a) $\lim_{n \to \infty} \|x_n - q\|$ exists for all $q \in F$

(b) $\lim_{n \to \infty} d(x_n, F)$ exists, where $d(x_n, F) = \inf_{p \in F} \|x_n - p\|$

Proof of Lemma C.1 1

Let $q \in F$, and set $l_n = \max\{l_n^{(1)}, l_n^{(2)}, \cdots, l_n^{(N)}\}$, $k_n = \max\{k_n^{(1)}, k_n^{(2)}, \cdots, k_n^{(N)}\}$. Since $\sum_{n=1}^{\infty} l_n^{(i)} < \infty$ and $\sum_{n=1}^{\infty} k_n^{(i)} < \infty$, it follows that $\sum_{n=1}^{\infty} l_n < \infty$ and $\sum_{n=1}^{\infty} k_n < \infty$. From how $\{x_n\}$ is generated, observe we have the following

$$\|y_n - q\| = \left\|P\left((1 - \beta_n)x_n + \beta_n T_i(PT_i)^{n-1}x_n\right) - Pq\right\|$$

$$\leq \|(1 - \beta_n)x_n + \beta_n T_i(PT_i)^{n-1}x_n - q\|$$

$$\leq (1 - \beta_n)\|x_n - q\| + \beta_n\|T_i(PT_i)^{n-1}x_n - q\|$$

$$\leq (1 - \beta_n)\|x_n - q\| + \beta_n(\frac{1}{2} + l_n)\left[\|I_i(PI_i)^{n-1}x_n - q\| + \|q - T_i(PT_i)^{n-1}x_n\|\right]$$

$$\leq (1 - \beta_n)\|x_n - q\| + 2\beta_n(1 + l_n)\|I_i(PI_i)^{n-1}x_n - q\|$$

$$\leq (1 - \beta_n)\|x_n - q\| + 2\beta_n(1 + l_n)(\frac{1}{2} + k_n)\left[\|x_n - q\| + \|q - I_i(PI_i)^{n-1}x_n\|\right]$$

$$\leq (1 - \beta_n)\|x_n - q\| + 4\beta_n(1 + l_n)(1 + k_n)\|x_n - q\|$$

$$\leq [1 + 3\beta_n + 4\beta_n k_n + 4\beta_n l_n + 4\beta_n l_n k_n]\|x_n - q\|$$

$$\leq [4 + 4k_n + 4l_n + 4l_n k_n]\|x_n - q\|$$

$$= 4(1 + l_n)(1 + k_n)\|x_n - q\|$$

On the other hand observe we have the following

$$\|x_{n+1} - q\| = \left\|P\left((1 - \alpha_n)T_i(PT_i)^{n-1}x_n + \alpha_n I_i(PI_i)^{n-1}y_n\right) - Pq\right\|$$

$$\leq \left\|(1 - \alpha_n)T_i(PT_i)^{n-1}x_n + \alpha_n I_i(PI_i)^{n-1}y_n\right) - q\right\|$$

$$\leq (1 - \alpha_n)\|T_i(PT_i)^{n-1}x_n - q\| + \alpha_n\|I_i(PI_i)^{n-1}y_n - q\|$$

$$\leq 2(1 - \alpha_n)(1 + l_n)\|I_i(PI_i)^{n-1}x_n - q\| + 2\alpha_n(1 + k_n)\|y_n - q\|$$

$$\leq 4(1 - \alpha_n)(1 + l_n)(1 + k_n)\|x_n - q\| + 8\alpha_n(1 + k_n)(1 + l_n)(1 + k_n)\|x_n - q\|$$

$$\leq [4(1 - \alpha_n) + 8\alpha_n(1 + k_n)](1 + l_n)(1 + k_n)\|x_n - q\|$$

$$\leq 8(1 + k_n)^2(1 + l_n)\|x_n - q\|$$

On the other hand since $\frac{1}{8}\|x_{n+1} - q\| < \|x_{n+1} - q\|$, then from the chain of inequalities immediately above we deduce that

$$\|x_{n+1} - q\| \leq (1 + k_n)^2(1 + l_n)\|x_n - q\|$$

$$\leq (1 + l_n + 2k_n + 2k_n l_n + k_n^2 + k_n^2 l_n)\|x_n - q\|$$

$$= (1 + \delta_n)\|x_n - q\|$$

where $\delta_n = l_n + 2k_n + 2k_n l_n + k_n^2 + k_n^2 l_n$. Since $\sum_{n=1}^{\infty} l_n < \infty$ and $\sum_{n=1}^{\infty} k_n < \infty$, it follows that $\sum_{n=1}^{\infty} \delta_n < \infty$. Now applying Lemma C.19 to the chain of inequalities immediately above we deduce that $\lim_{n\to\infty} \|x_n - q\|$ exists, and so part (a) of the theorem follows. To see part (b), take infimum over all $q \in F$, in the chain of inequalities immediately above, we deduce that

$$d(x_{n+1}, F) \leq (1 + \delta_n)d(x_n, F)$$

Now applying Lemma C.19 to the inequality immediately above, it follows that $\lim_{n\to\infty} d(x_n, F)$ exists

Lemma C.2 1

Let X be a real uniformly convex Banach space, K be a nonempty closed convex subset of X which is also a $(\delta, 1 - \delta)$ nonexpansive retract with retraction P. Let $\{T_i\}_i^N : K \mapsto X$ be N nonself I_i-asymptotically $(\delta, 1 - \delta)$ nonexpansive mappings with sequences $\{l_n^{(i)}\} \subset [0, \infty)$ such that $\sum_{n=1}^{\infty} l_n^{(i)} < \infty$ and $\{I_i\}_i^N : K \mapsto X$ be N nonself asymptotically $(\delta, 1 - \delta)$ nonexpansive mappings with sequences $\{k_n^{(i)}\} \subset [0, \infty)$ such that $\sum_{n=1}^{\infty} k_n^{(i)} < \infty$. Let $\{\alpha_n\}$ and $\{\beta_n\}$ be sequences in $[a, 1 - a]$ for some $a \in (0, 1)$. Suppose for any given $x_1 \in K$, the sequence $\{x_n\}$ is generated as follows for all $n \geq 1$

$$x_{n+1} = P\left((1 - \alpha_n)T_i(PT_i)^{n-1}x_n + \alpha_n I_i(PI_i)^{n-1}y_n\right)$$

$$y_n = P\left((1 - \beta_n)x_n + \beta_n T_i(PT_i)^{n-1}x_n\right)$$

and $F = \bigcap_{i=1}^{N} F(T_i) \cap F(I_i) \neq \emptyset$. Then

$$\lim_{n \to \infty} \|x_n - T_i x_n\| = \lim_{n \to \infty} \|x_n - I_i x_n\| = 0$$

for all $i \in J$

Proof of Lemma C.2 1

By the previous Lemma, $\lim_{n \to \infty} \|x_n - q\|$ exists for all $q \in F$. Call it c. From the previous theorem, we know that

$$\|y_n - q\| \le 4(1 + l_n)(1 + k_n)\|x_n - q\|$$

Now taking $\lim \sup$ on both sides in the equality immediately above, we deduce that $\lim \sup_{n \to \infty} \|y_n - q\| \le \lim \sup_{n \to \infty} \|x_n - q\| = \lim_{n \to \infty} \|x_n - q\| = c$. It follows that

$$\|T_i(PT_i)^{n-1}x_n - q\| \le 4(1 + l_n)(1 + k_n)\|x_n - q\|$$

for all $n \ge 1$ implies that

$$\lim_{n \to \infty} \sup \|T_i(PT_i)^{n-1}x_n - q\| \le c$$

Since $\lim \sup_{n \to \infty} \|y_n - q\| \le c$ and $\|I_i(PI_i)^{n-1}y_n - q\| \le 2(1 + k_n)\|y_n - q\|$ we deduce the following

$$\lim_{n \to \infty} \sup \|I_i(PI_i)^{n-1}y_n - q\| \le c$$

From the proof of the previous Lemma, we know that,

$$\|x_{n+1} - q\| \le \|(1 - \alpha_n)(T_i(PT_i)^{n-1}x_n - q) + \alpha_n(I_i(PI_i)^{n-1}y_n - q)\|$$
$$\le (1 + \delta_n)\|x_n - q\|$$

Since $\sum_{n=1}^{\infty} \delta_n < \infty$ and $\lim_{n \to \infty} \|x_{n+1} - q\| = c$, taking limits in the inequality immediately above we deduce that

$$\lim_{n \to \infty} \|(1 - \alpha_n)(T_i(PT_i)^{n-1}x_n - q) + \alpha_n(I_i(PI_i)^{n-1}y_n - q)\| = c$$

By the above equality, and the fact that $\lim \sup_{n \to \infty} \|T_i(PT_i)^{n-1}x_n - q\| \le c$, and

$$\lim_{n \to \infty} \sup \|I_i(PI_i)^{n-1}y_n - q\| \le c$$

then upon using Lemma C.20, we deduce the following

$$\lim_{n \to \infty} \|T_i(PT_i)^{n-1}x_n - I_i(PI_i)^{n-1}y_n\| = 0$$

From how $\{x_n\}$ is generated, we deduce the following

$$\|x_{n+1} - q\| \le \|(1 - \alpha_n)T_i(PT_i)^{n-1}x_n + \alpha_n I_i(PI_i)^{n-1}y_n - q\|$$
$$\le \|T_i(PT_i)^{n-1}x_n - q\| + \alpha_n\|T_i(PT_i)^{n-1}x_n - I_i(PI_i)^{n-1}y_n\|$$

Since $\lim_{n \to \infty} \|T_i(PT_i)^{n-1}x_n - I_i(PI_i)^{n-1}y_n\| = 0$, and $\lim_{n \to \infty} \|x_{n+1} - q\| = c$, if we take $\lim \inf$ on both sides in the inequality immediately above, we deduce the following

$$c \le \lim_{n \to \infty} \inf \|T_i(PT_i)^{n-1}x_n - q\|$$

Since $\lim \sup_{n \to \infty} \|T_i(PT_i)^{n-1}x_n - q\| \le c$, then combining with inequality immediately above, we deduce the following

$$\lim_{n \to \infty} \|T_i(PT_i)^{n-1}x_n - q\| = c$$

Proof of Lemma C.2 Continued 1

Now observe we have the following

$$\|T_i(PT_i)^{n-1}x_n - q\| \le \|T_i(PT_i)^{n-1}x_n - I_i(PI_i)^{n-1}y_n\| + \|I_i(PI_i)^{n-1}y_n - q\|$$
$$\le \|T_i(PT_i)^{n-1}x_n - I_i(PI_i)^{n-1}y_n\| + 2(1+k_n)\|y_n - q\|$$

Since $\lim_{n\to\infty}\|T_i(PT_i)^{n-1}x_n - q\| = c$ and $\lim_{n\to\infty}\|T_i(PT_i)^{n-1}x_n - I_i(PI_i)^{n-1}y_n\| = 0$, if we take $\liminf$ on both sides in the inequality immediately above, we deduce the following

$$c \le \lim\inf_{n\to\infty}\|y_n - q\|$$

Combining the above inequality with $\limsup_{n\to\infty}\|y_n - q\| \le c$, we obtain that

$$\lim_{n\to\infty}\|y_n - q\| = c$$

From the proof of the previous Lemma, we deduce the following

$$\|y_n - q\| \le \|(1-\beta_n)(x_n - q) + \beta_n(T_i(PT_i)^{n-1}x_n - q)\|$$
$$\le 4(1+l_n)(1+k_n)\|x_n - q\|$$

Since $\lim_{n\to\infty}\|y_n-q\| = c$ and $\lim_{n\to\infty}\|x_n-q\| = c$, if we take limits in the above inequality we deduce the following

$$\lim_{n\to\infty}\|(1-\beta_n)(x_n - q) + \beta_n(T_i(PT_i)^{n-1}x_n - q)\| = c$$

Since $\lim_{n\to\infty}\|x_n - q\| = c$ and $\lim_{n\to\infty}\|T_i(PT_i)^{n-1}x_n - q\| = c$, then from the above equality and Lemma 1.20, we deduce that

$$\lim_{n\to\infty}\|x_n - T_i(PT_i)^{n-1}x_n\| = 0$$

From the above inequality, $y_n = P\left((1-\beta_n)x_n + \beta_nT_i(PT_i)^{n-1}x_n\right)$, and the $(\delta, 1-\delta)$ non-expansiveness of P, we deduce the following

$$\|y_n - x_n\| = \|P\left((1-\beta_n)x_n + \beta_nT_i(PT_i)^{n-1}x_n\right) - Px_n\|$$
$$\le \|(1-\beta_n)x_n + \beta_nT_i(PT_i)^{n-1}x_n - x_n\|$$
$$\le \beta_n\|T_i(PT_i)^{n-1}x_n - x_n\|$$
$$\to 0 \ (n \to \infty)$$

Now from $\lim_{n\to\infty}\|x_n - T_i(PT_i)^{n-1}x_n\| = 0$, $x_{n+1} = P\left((1-\alpha_n)T_i(PT_i)^{n-1}x_n + \alpha_nI_i(PI_i)^{n-1}y_n\right)$, and the $(\delta, 1-\delta)$ non-expansiveness of P, we deduce the following

$$\|x_{n+1} - x_n\| \le \|P\left((1-\alpha_n)T_i(PT_i)^{n-1}x_n + \alpha_nI_i(PI_i)^{n-1}y_n\right) - Px_n\|$$
$$\le \|(1-\alpha_n)T_i(PT_i)^{n-1}x_n + \alpha_nI_i(PI_i)^{n-1}y_n - x_n\|$$
$$\le \|T_i(PT_i)^{n-1}x_n - x_n\| + \alpha_n\|T_i(PT_i)^{n-1}x_n - I_i(PI_i)^{n-1}y_n\|$$
$$\to 0 \ (n \to \infty)$$

From $\lim_{n\to\infty}\|y_n - x_n\| = 0$ and $\lim_{n\to\infty}\|x_{n+1} - x_n\| = 0$, we deduce that

$$\lim_{n\to\infty}\|x_{n+1} - y_n\| = 0$$

Proof of Lemma C.2 Continued 1

Now observe that

$$\|x_n - I_i(PI_i)^{n-1}y_n\| \leq \|x_n - T_i(PT_i)^{n-1}x_n\| + \|T_i(PT_i)^{n-1}x_n - I_i(PI_i)^{n-1}y_n\|$$

Since $\lim_{n \to \infty} \|x_n - T_i(PT_i)^{n-1}x_n\| = 0$ and $\lim_{n \to \infty} \|T_i(PT_i)^{n-1}x_n - I_i(PI_i)^{n-1}y_n\| = 0$, then taking limits in the above inequality, we deduce that

$$\lim_{n \to \infty} \|x_n - I_i(PI_i)^{n-1}y_n\| = 0$$

Also

$$\|x_{n+1} - I_i(PI_i)^{n-1}y_n\| \leq \|x_{n+1} - x_n\| + \|x_n - I_i(PI_i)^{n-1}y_n\|$$

Since $\lim_{n \to \infty} \|x_{n+1} - x_n\| = 0$, and $\lim_{n \to \infty} \|x_n - I_i(PI_i)^{n-1}y_n\| = 0$, then taking limits in the above inequality, we deduce that

$$\lim_{n \to \infty} \|x_{n+1} - I_i(PI_i)^{n-1}y_n\| = 0$$

Now observe we have the following

$$\begin{aligned}
\|x_n - I_i(PI_i)^{n-1}x_n\| &\leq \|x_n - x_{n+1}\| + \|x_{n+1} - I_i(PI_i)^{n-1}y_n\| \\
&\quad + \|I_i(PI_i)^{n-1}y_n - I_i(PI_i)^{n-1}x_n\| \\
&\leq \|x_n - x_{n+1}\| + \|x_{n+1} - I_i(PI_i)^{n-1}y_n\| + 2(1 + k_n)\|y_n - x_n\|
\end{aligned}$$

Since $\lim_{n \to \infty} \|x_n - x_{n+1}\| = \lim_{n \to \infty} \|x_{n+1} - I_i(PI_i)^{n-1}y_n\| = \lim_{n \to \infty} \|y_n - x_n\| = 0$, then taking limits in the inequality immediately above, we deduce that

$$\lim_{n \to \infty} \|x_n - I_i(PI_i)^{n-1}x_n\| = 0$$

Now since an asymptotically $(\delta, 1 - \delta)$ nonexpansive mapping with respect to P must be uniformly Lipschitzian with respect to P, then we have the following

$$\begin{aligned}
\|x_{n+1} - I_i x_{n+1}\| &\leq \|x_{n+1} - I_i(PI_i)^n x_{n+1}\| + \|I_i(PI_i)^n x_{n+1} - I_i x_{n+1}\| \\
&\leq \|x_{n+1} - I_i(PI_i)^n x_{n+1}\| + L\|I_i(PI_i)^{n-1}x_{n+1} - x_{n+1}\| \\
&\quad + L\|x_{n+1} - I_i(PI_i)^n x_{n+1}\| \\
&\leq \|x_{n+1} - I_i(PI_i)^n x_{n+1}\| + 2L\|I_i(PI_i)^{n-1}x_{n+1} - x_{n+1}\| \\
&\leq \|x_{n+1} - I_i(PI_i)^n x_{n+1}\| + 2L\|I_i(PI_i)^{n-1}x_{n+1} - I_i(PI_i)^{n-1}y_n\| \\
&\quad + 2L\|I_i(PI_i)^{n-1}y_n - x_{n+1}\| \\
&\leq \|x_{n+1} - I_i(PI_i)^n x_{n+1}\| + 2L\|I_i(PI_i)^{n-1}y_n - x_{n+1}\| \\
&\quad + 2L^2\big[\|x_{n+1} - y_n\| + \|y_n - I_i(PI_i)^{n-1}x_{n+1}\|\big] \\
&\leq \|x_{n+1} - I_i(PI_i)^n x_{n+1}\| + 2L\|I_i(PI_i)^{n-1}y_n - x_{n+1}\| \\
&\quad + 4L^2\|x_{n+1} - y_n\|
\end{aligned}$$

Since $\lim_{n \to \infty} \|x_{n+1} - I_i(PI_i)^n x_{n+1}\| = \lim_{n \to \infty} \|I_i(PI_i)^{n-1}y_n - x_{n+1}\| = \lim_{n \to \infty} \|x_{n+1} - y_n\| = 0$, it follows that

$$\lim_{n \to \infty} \|x_n - I_i x_n\| = 0$$

Proof of Lemma C.2 Continued 1

Since every nonself I-asymptotically $(\delta, 1 - \delta)$ nonexpansive mapping with respect to P must be I-uniformly Lipschitz with respect to P, then we have the following

$$
\begin{aligned}
\|x_{n+1} - T_i x_{n+1}\| &\leq \|x_{n+1} - T_i(PT_i)^n x_{n+1}\| + \|T_i(PT_i)^n x_{n+1} - T_i(PT_i)^n x_n\| \\
&\quad + \|T_i(PT_i)^n x_n - T_i x_{n+1}\| \\
&\leq \|x_{n+1} - T_i(PT_i)^n x_{n+1}\| + \Gamma \Big[\|I_i(PI_i)^n x_{n+1} - I_i(PI_i)^n x_n\| \\
&\quad + \|I_i(PI_i)^n x_n - T_i(PT_i)^n x_{n+1}\| \Big] + \Gamma \Big[\|I_i(PI_i)^n x_n - I_i x_{n+1}\| \\
&\quad + \|I_i x_{n+1} - T_i(PT_i)^n x_n\| \Big] \\
&\leq \|x_{n+1} - T_i(PT_i)^n x_{n+1}\| + 2\Gamma \|I_i(PI_i)^n x_{n+1} - I_i(PI_i)^n x_n\| \\
&\quad + 2\Gamma \|I_i(PI_i)^n x_n - I_i x_{n+1}\| \\
&\leq \|x_{n+1} - T_i(PT_i)^n x_{n+1}\| + 4L\Gamma \|x_{n+1} - x_n\| + 4L\Gamma \|I_i(PI_i)^{n-1} x_n - x_{n+1}\| \\
&\leq \|x_{n+1} - T_i(PT_i)^n x_{n+1}\| + 4L\Gamma \|x_{n+1} - x_n\| + 4L\Gamma \|I_i(PI_i)^{n-1} x_n - x_n\| \\
&\quad + 4L\Gamma \|x_n - x_{n+1}\|
\end{aligned}
$$

Since as $n \to \infty$, $\|x_{n+1} - T_i(PT_i)^n x_{n+1}\|, \|x_{n+1} - x_n\|, \|I_i(PI_i)^{n-1} x_n - x_n\| \to 0$, it follows from the inequality immediately above, that, $\lim_{n \to \infty} \|x_{n+1} - T_i x_{n+1}\| = 0$ and the proof is complete

Now our strong convergence theorem in Banach space is as follows

Theorem C.3 1

Let X be a real Banach space, K be a nonempty closed convex subset of X which is also a $(\delta, 1 - \delta)$ nonexpansive retract with retraction P. Let $\{T_i\}_i^N : K \mapsto X$ be N nonself I_i-asymptotically $(\delta, 1 - \delta)$ nonexpansive mappings with sequences $\{l_n^{(i)}\} \subset [0, \infty)$ such that $\sum_{n=1}^{\infty} l_n^{(i)} < \infty$ and $\{I_i\}_i^N : K \mapsto X$ be N nonself asymptotically $(\delta, 1 - \delta)$ nonexpansive mappings with sequences $\{k_n^{(i)}\} \subset [0, \infty)$ such that $\sum_{n=1}^{\infty} k_n^{(i)} < \infty$. Suppose for any given $x_1 \in K$, the sequence $\{x_n\}$ is generated as follows for all $n \geq 1$

$$
x_{n+1} = P\left((1 - \alpha_n) T_i(PT_i)^{n-1} x_n + \alpha_n I_i(PI_i)^{n-1} y_n \right)
$$

$$
y_n = P\left((1 - \beta_n) x_n + \beta_n T_i(PT_i)^{n-1} x_n \right)
$$

and $F = \bigcap_{i=1}^{N} F(T_i) \cap F(I_i) \neq \emptyset$. Then $\{x_n\}$ converges strongly to a common fixed point of $\{T_i\}_i^N$ and $\{I_i\}_i^N$ iff $\liminf_{n \to \infty} d(x_n, F) = 0$

Proof of Theorem C.3 1

The necessity portion of the proof is obvious. For sufficiency, let $q \in F$, then by Lemma C.1(b), $\lim_{n \to \infty} d(x_n, F)$ exists, and by assumption $\liminf_{n \to \infty} d(x_n, F) = 0$, thus, $\lim_{n \to \infty} d(x_n, F) = 0$. Now we show that $\{x_n\}$ is a Cauchy sequence in K. Note from the proof of Lemma C.1 that for any $q \in K$, we have the following

$$\|x_{n+m} - q\| \leq \exp\left(\sum_{n=1}^{\infty} \delta_n\right) \|x_n - q\|$$

$$< M\|x_n - q\|$$

for all m, n, where $M = \exp\left(\sum_{n=1}^{\infty} \delta_n\right) + 1$. Since $\lim_{n \to \infty} d(x_n, F) = 0$, for any given $\epsilon > 0$, there exists a positive integer N_0 such that for all $n \geq N_0$, $d(x_n, F) < \frac{\epsilon}{2M}$. Also there exists $q_1 \in F$ such that $\|x_{n_0} - q_1\| < \frac{\epsilon}{2M}$. Now for all $n \geq N_0$ and $m \geq 1$, we have the following

$$\|x_{n+m} - x_n\| \leq \|x_{n+m} - q_1\| + \|x_n - q_1\|$$

$$\leq M\|x_{n_0} - q_1\| + M\|x_{n_0} - q_1\|$$

$$\leq 2M\|x_{n_0} - q_1\|$$

$$\leq 2M\frac{\epsilon}{2M}$$

$$= \epsilon$$

which implies that $\{x_n\}$ is a Cauchy sequence in K. By the completeness of X, $\{x_n\}$ is convergent. Assume that $\{x_n\}$ converges to a point q. Then $q \in K$, because K is a closed subset of X. Now $\lim_{n \to \infty} d(x_n, F) = 0$ implies that $\lim_{n \to \infty} d(q, F) = 0$. Since F is closed, $q \in F$, and the proof is finished.

If in Definition C.18, $\{T_i\}_i^N : K \mapsto X$ are N nonself I_i-asymptotically $(\delta, 1 - \delta)$ nonexpansive mappings and $\{I_i\}_i^N : K \mapsto X$ are N nonself asymptotically $(\delta, 1 - \delta)$ nonexpansive mappings, then upon applying the previous theorem we get the following strong convergence theorem in real uniformly convex Banach space as follows

Theorem C.4 1

Let X be a real uniformly convex Banach space, K be a nonempty closed convex subset of X which is also a $(\delta, 1 - \delta)$ nonexpansive retract with retraction P. Let $\{T_i\}_i^N : K \mapsto X$ be N nonself I_i-asymptotically $(\delta, 1 - \delta)$ nonexpansive mappings with sequences $\{l_n^{(i)}\} \subset [0, \infty)$ such that $\sum_{n=1}^{\infty} l_n^{(i)} < \infty$ and $\{I_i\}_i^N : K \mapsto X$ be N nonself asymptotically $(\delta, 1 - \delta)$ nonexpansive mappings with sequences $\{k_n^{(i)}\} \subset [0, \infty)$ such that $\sum_{n=1}^{\infty} k_n^{(i)} < \infty$. Let $\{\alpha_n\}$ and $\{\beta_n\}$ be sequences in $[a, 1 - a]$ for some $a \in (0, 1)$. Suppose for any given $x_1 \in K$, the sequence $\{x_n\}$ is generated as follows for all $n \geq 1$

$$x_{n+1} = P\left((1 - \alpha_n)T_i(PT_i)^{n-1}x_n + \alpha_n I_i(PI_i)^{n-1}y_n\right)$$

$$y_n = P\left((1 - \beta_n)x_n + \beta_n T_i(PT_i)^{n-1}x_n\right)$$

and $F = \bigcap_{i=1}^{N} F(T_i) \cap F(I_i) \neq \emptyset$. If $\{T_i\}_i^N$ and $\{I_i\}_i^N$ satisfy Definition C.18, then $\{x_n\}$ converges strongly to a common fixed point of $\{T_i\}_i^N$ and $\{I_i\}_i^N$

Proof of Theorem C.4 1

Since $\{T_i\}_i^N$ and $\{I_i\}_i^N$ satisfy Definition C.18, it follows that

$$\max_{1 \leq i \leq N} \left[\frac{1}{2} [\|x_n - T_i x_n\| + \|x_n - I_i x_n\|] \right] \geq f(d(x_n, F))$$

From Lemma C.2, we know that $\lim_{n \to \infty} \|x_n - T_i x_n\| = \lim_{n \to \infty} \|x_n - I_i x_n\| = 0$ for all $i \in J$, thus taking limits in the above inequality we deduce the following $\lim_{n \to \infty} f(d(x_n, F)) = 0$. Since $f : [0, \infty) \mapsto [0, \infty)$ is a nondecreasing function satisfying $f(0) = 0$, and $f(r) > 0$ for all $r \in (0, \infty)$, it follows that $\lim_{n \to \infty} d(x_n, F) = 0$. Now all the conditions of the previous theorem are satisfied, therefore by its conclusion $\{x_n\}$ converges strongly to a point of F

3.4 Exercises

Recall from [Clement Boateng Ampadu, A Strong Ciric Almost Contraction Mapping Theorem in Metric Spaces with Unique Fixed Point, Applied Mathematics(AM), Initially Accepted but Withdrawn] that if (X, d) is a metric space, then a map $T : X \mapsto X$ is called a strong Ciric $(\alpha, 1 - \alpha)$-weak contraction if there exists $\alpha \in (0, 1)$ such that for all $x, y \in X$ the following holds

$$d(Tx, Ty) \leq \alpha M_1(x, y) + (1 - \alpha) d(y, Tx)$$

where

$$M_1(x, y) = \max\{d(x, y), d(x, Tx), d(y, Ty), \frac{1}{2}[d(x, Ty) + d(y, Tx)]\}$$

Observe

$$d(Tx, Ty) \leq \alpha M_1(x, y) + (1 - \alpha) d(y, Tx)$$

where

$$M_1(x, y) = \max\{d(x, y), d(x, Tx), d(y, Ty), \frac{1}{2}[d(x, Ty) + d(y, Tx)]\}$$

is equivalent to

$$d(Tx, Ty) \leq (\alpha + 1 - \alpha) \max\{M_1(x, y), d(y, Tx)\}$$

where

$$M_1(x, y) = \max\{d(x, y), d(x, Tx), d(y, Ty), \frac{1}{2}[d(x, Ty) + d(y, Tx)]\}$$

Now let K be a nonempty subset of a real normed space X and $P : X \mapsto K$ be a $(\alpha, 1 - \alpha)$-nonexpansive retraction of X onto K

(a) $T : K \mapsto X$ will be called asymptotically $(\alpha, 1 - \alpha)$ nonexpansive if there exists a sequence $\{k_n\} \in [0, \infty)$, with $\lim_{n \to \infty} k_n = 0$, such that

$$\|T(PT)^{n-1}x - T(PT)^{n-1}y\| \leq (\frac{1}{2} + k_n) \left[J(x, y) + \|y - T(PT)^{n-1}x\| \right]$$

holds for all $x, y \in K$ and any positive integer n, where

$$J(x, y) = \max\{\|x - y\|, \|x - T(PT)^{n-1}x\|, \|y - T(PT)^{n-1}y\|, \frac{1}{2}[\|x - T(PT)^{n-1}y\| + \|y - T(PT)^{n-1}x\|]\}$$

(b) If $T, I : K \mapsto X$ are two non-self mappings. $T : K \mapsto X$ will be called I-asymptotically $(\delta, 1 - \delta)$ nonexpansive if there exists a sequence $\{u_n\} \in [0, \infty)$, with $\lim_{n \to \infty} u_n = 0$, such that

$$\|T(PT)^{n-1}x - T(PT)^{n-1}y\| \leq (\frac{1}{2} + u_n) \left[J_1(x, y) + \|I(PI)^{n-1}y - T(PT)^{n-1}x\| \right]$$

holds for all $x, y \in K$ and any positive integer n, where

$$J_1(x,y) = \max\{\|I(PI)^{n-1}x - I(PI)^{n-1}y\|, \|I(PI)^{n-1}x - T(PT)^{n-1}x\|, \|I(PI)^{n-1}y - T(PT)^{n-1}y\|,$$

$$\frac{1}{2}[\|I(PI)^{n-1}x - T(PT)^{n-1}y\| + \|I(PI)^{n-1}y - T(PT)^{n-1}x\|]\}$$

Exercise C.1 1

Let X be a real Banach space, K be a nonempty closed convex subset of X which is also a $(\alpha, 1-\alpha)$ nonexpansive retract with retraction P. Let $\{T_i\}_i^N : K \mapsto X$ be N nonself I_i-asymptotically $(\alpha, 1-\alpha)$ nonexpansive mappings with sequences $\{l_n^{(i)}\} \subset [0,\infty)$ such that $\sum_{n=1}^{\infty} l_n^{(i)} < \infty$ and $\{I_i\}_i^N : K \mapsto X$ be N nonself asymptotically $(\alpha, 1-\alpha)$ nonexpansive mappings with sequences $\{k_n^{(i)}\} \subset [0,\infty)$ such that $\sum_{n=1}^{\infty} k_n^{(i)} < \infty$. Suppose that for any given $x_1 \in K$, the sequence $\{x_n\}$ is generated as follows for all $n \geq 1$

$$x_{n+1} = P\left((1-\alpha_n)T_i(PT_i)^{n-1}x_n + \alpha_n I_i(PI_i)^{n-1}y_n\right)$$

$$y_n = P\left((1-\beta_n)x_n + \beta_n T_i(PT_i)^{n-1}x_n\right)$$

and $F = \bigcap_{i=1}^{N} F(T_i) \cap F(I_i) \neq \emptyset$. Then

(a) $\lim_{n\to\infty} \|x_n - q\|$ exists for all $q \in F$

(b) $\lim_{n\to\infty} d(x_n, F)$ exists, where $d(x_n, F) = \inf_{p \in F} \|x_n - p\|$

Exercise C.2 1

Let X be a real uniformly convex Banach space, K be a nonempty closed convex subset of X which is also a $(\alpha, 1-\alpha)$ nonexpansive retract with retraction P. Let $\{T_i\}_i^N : K \mapsto X$ be N nonself I_i-asymptotically $(\alpha, 1-\alpha)$ nonexpansive mappings with sequences $\{l_n^{(i)}\} \subset [0,\infty)$ such that $\sum_{n=1}^{\infty} l_n^{(i)} < \infty$ and $\{I_i\}_i^N : K \mapsto X$ be N nonself asymptotically $(\alpha, 1-\alpha)$ nonexpansive mappings with sequences $\{k_n^{(i)}\} \subset [0,\infty)$ such that $\sum_{n=1}^{\infty} k_n^{(i)} < \infty$. Let $\{\alpha_n\}$ and $\{\beta_n\}$ be sequences in $[a, 1-a]$ for some $a \in (0,1)$. Suppose for any given $x_1 \in K$, the sequence $\{x_n\}$ is generated as follows for all $n \geq 1$

$$x_{n+1} = P\left((1-\alpha_n)T_i(PT_i)^{n-1}x_n + \alpha_n I_i(PI_i)^{n-1}y_n\right)$$

$$y_n = P\left((1-\beta_n)x_n + \beta_n T_i(PT_i)^{n-1}x_n\right)$$

and $F = \bigcap_{i=1}^{N} F(T_i) \cap F(I_i) \neq \emptyset$. Then

$$\lim_{n\to\infty} \|x_n - T_i x_n\| = \lim_{n\to\infty} \|x_n - I_i x_n\| = 0$$

for all $i \in J$

Exercise C.3 1

Let X be a real Banach space, K be a nonempty closed convex subset of X which is also a $(\alpha, 1 - \alpha)$ nonexpansive retract with retraction P. Let $\{T_i\}_i^N : K \mapsto X$ be N nonself I_i-asymptotically $(\alpha, 1 - \alpha)$ nonexpansive mappings with sequences $\{l_n^{(i)}\} \subset [0, \infty)$ such that $\sum_{n=1}^{\infty} l_n^{(i)} < \infty$ and $\{I_i\}_i^N : K \mapsto X$ be N nonself asymptotically $(\alpha, 1 - \alpha)$ nonexpansive mappings with sequences $\{k_n^{(i)}\} \subset [0, \infty)$ such that $\sum_{n=1}^{\infty} k_n^{(i)} < \infty$. Suppose for any given $x_1 \in K$, the sequence $\{x_n\}$ is generated as follows for all $n \geq 1$

$$x_{n+1} = P\bigg((1 - \alpha_n)T_i(PT_i)^{n-1}x_n + \alpha_n I_i(PI_i)^{n-1}y_n\bigg)$$

$$y_n = P\bigg((1 - \beta_n)x_n + \beta_n T_i(PT_i)^{n-1}x_n\bigg)$$

and $F = \bigcap_{i=1}^{N} F(T_i) \cap F(I_i) \neq \emptyset$. Then $\{x_n\}$ converges strongly to a common fixed point of $\{T_i\}_i^N$ and $\{I_i\}_i^N$ iff $\liminf_{n \to \infty} d(x_n, F) = 0$

Exercise C.4 1

If in Definition C.18, $\{T_i\}_i^N : K \mapsto X$ are N nonself I_i-asymptotically $(\alpha, 1 - \alpha)$ nonexpansive mappings and $\{I_i\}_i^N : K \mapsto X$ are N nonself asymptotically $(\alpha, 1 - \alpha)$ nonexpansive mappings, then upon applying the previous exercise, prove that we get the following strong convergence theorem in real uniformly convex Banach space:

Let X be a real uniformly convex Banach space, K be a nonempty closed convex subset of X which is also a $(\alpha, 1 - \alpha)$ nonexpansive retract with retraction P. Let $\{T_i\}_i^N : K \mapsto X$ be N nonself I_i-asymptotically $(\alpha, 1 - \alpha)$ nonexpansive mappings with sequences $\{l_n^{(i)}\} \subset [0, \infty)$ such that $\sum_{n=1}^{\infty} l_n^{(i)} < \infty$ and $\{I_i\}_i^N : K \mapsto X$ be N nonself asymptotically $(\alpha, 1 - \alpha)$ nonexpansive mappings with sequences $\{k_n^{(i)}\} \subset [0, \infty)$ such that $\sum_{n=1}^{\infty} k_n^{(i)} < \infty$. Let $\{\alpha_n\}$ and $\{\beta_n\}$ be sequences in $[a, 1 - a]$ for some $a \in (0, 1)$. Suppose for any given $x_1 \in K$, the sequence $\{x_n\}$ is generated as follows for all $n \geq 1$

$$x_{n+1} = P\bigg((1 - \alpha_n)T_i(PT_i)^{n-1}x_n + \alpha_n I_i(PI_i)^{n-1}y_n\bigg)$$

$$y_n = P\bigg((1 - \beta_n)x_n + \beta_n T_i(PT_i)^{n-1}x_n\bigg)$$

and $F = \bigcap_{i=1}^{N} F(T_i) \cap F(I_i) \neq \emptyset$. If $\{T_i\}_i^N$ and $\{I_i\}_i^N$ satisfy Definition C.18, then $\{x_n\}$ converges strongly to a common fixed point of $\{T_i\}_i^N$ and $\{I_i\}_i^N$

3.5 References

(1) Birol Gündüz and Sezgin Akbulut, On Weak and Strong Convergence Theorems for a Finite Family of Nonself I-asymptotically Nonexpansive Mappings, Mathematica Moravica Vol. 19-2 (2015), 49–64

(2) Clement Boateng Ampadu, An Almost Contraction Mapping Theorem in Metric Spaces with Unique Fixed Point, Submitted. Available online:
https://drive.google.com/file/d/0BwtkpMtWoUlEY25OZW1HUEdGcU0/view

(3) K. Goebel, W.A. Kirk, A fixed point theorem for asymptotically nonexpansive mappings, Proc. Amer. Math. Soc., 35 (1972), 171–174

(4) C.E. Chidume, E.U. Ofoedu, H. Zegeye, Strong and weak convergence theorems for asymptotically nonexpansive mappings, J. Math. Anal. Appl., 280 (2003), 364-374

(5) S. Temir, On the convergence theorems of implicit iteration process for a finite family of I-asymptotically nonexpansive mappings, J. Comput. Appl. Math. 225 (2009),398–405

(6) S. Temir, O. Gul, Convergence theorem for I-asymptotically quasi-nonexpansive mapping in Hilbert space, J. Math. Anal. Appl. 329 (2007), 759–765

(7) L.Yang, X. Xie, Weak and strong convergence theorems for a finite family of I-asymptotically nonexpansive mappings, Appl. Math. Comput. 216 (2010), 1057–1064

(8) M.O. Osilike, A. Udomene, Demiclosedness principle and convergence theorems for strictly pseudocontractive mappings of Browder–Petryshyn type, J. Math. Anal. Appl. 256 (2001), 431–445

(9) H.F. Senter, W.G. Dotson, Approximating fixed points of nonexpansive mappings, Proc. Amer. Math. Soc., 44 (1974), 375-380

(10) S.H. Khan, H. Fukharuddin, Weak and strong convergence of a scheme with errors for two nonexpansive mappings, Nonlinear Anal. 61 (2005), 1295–1301

(11) K.K. Tan and H.K. Xu, Approximating fixed points of nonexpansive mappings by the Ishikawa iteration process, Journal of Mathematical Analysis and Applications, vol. 178, no. 2 (1993), pp. 301–308

(12) J. Schu, Weak and strong convergence to fixed points of asymptotically nonexpansive mappings, Bull. Austral. Math. Soc. 43 (1991), 153-159

(13) Clement Boateng Ampadu, A Strong Ciric Almost Contraction Mapping Theorem in Metric Spaces with Unique Fixed Point, Applied Mathematics(AM), Initially Accepted but Withdrawn. Available online:
https://drive.google.com/file/d/0BwtkpMtWoUlEZHhpdExKVGdMOTA/view